Dr Kesorn Pechrach Weaver

PIEZOELECTRICS
IN
CIRCUIT BREAKERS

Design & Test

Piezoelectrics
in
Circuit Breakers

Design & Test

Dr Kesorn Pechrach Weaver

Pechrach Publishing

Piezoelectrics in Circuit Breakers
Design & Test
By
Dr Kesorn Pechrach Weaver

ISBN 978-0-9931178-0-0

PECHRACH PUBLISHING
7 Boundary Road, Bishops Stortford, Hertfordshire, CM23 5LE, England, United Kingdom. Tel: (+44) 1279 508020, +44(0) 7779913907

This book is dedicated to my family.

Acknowledgments

I would like to thank my beautiful son, Neran J. P. Weaver for his non-stop supporting.

Many Thanks to my family in Thailand for always believe in me.

I would like to thank Prof. J.W. McBride for showing me a new way of life.

A special thanks to J. P. Darby for looking after my child while I write this book.

Table of Contents

Acknowledgement

Table of Figures

velocity: 10 m/s

and point "C" is the point at which the cathode root moves from fixed contact

List of Tables

Nomenclature

B   magnetic flux density

Permeability of free space ($4\pi \times 10^{-7}$ H/m)

γ specific heat ratio CP/CV = 1.4 (air)

r   magnitude of distance from current moment

to point S at which magnetic effect was evaluated

i   arc current

j   current density

a   contact gap

VA   Arc voltage

σ electrical conductivity

E   electric field

P0   pressure at the shock front

P1   pressure in front of the arc

T0   temperature in front of the shock wave

T1   temperature behind the shock wave

ao   sound speed

l   arc length

Fmag magnetic forces

Fg   gas dynamic forces

Darc arc diameter

Ps   stagnation pressure

M   Mach number

K   Nozzle flow constant gas

h   enthalpy of metal vapour (overheating energies, melting volatilization and ionization)

Ro   Ideal gases constant

cv   specific heat at constant volume

cp   specific heat at constant pressure

Ts   starting point temperature

TB   boiling point temperature

Tm   melting point temperature

TG   gas temperature

J   the total current density,

Ji   the ion current density,

Jem   the electron current density

Parc arc power

CHAPTER 1

CIRCUIT BREAKERS

Low voltage miniature circuit breakers (MCBs) are widely used for protection from electrical faults in domestic and light industrial installations. These devices protect against short circuit faults of typically 1–16 kA. During a short circuit fault an electric arc is drawn between opening contacts. The arc is then propelled by electromagnetic forces into splitter plates which divide the arc into a number of smaller arcs. This results in a high arc voltage, which counteracts the supply voltage to limit the peak fault current. The energy released by the fault is reduced and damage to both the circuit and the circuit breaker is minimized.

To achieve satisfactory short circuit performance rapid contact opening and a high contact velocity are usually necessary.

At low contact opening velocity, the arc root stays on the contacts for a longer period of time. Heat is generated in the arc chamber and the contacts are eroded. This has resulted in a range of devices where the opening mechanism is driven by a solenoid energised by the short circuit current. The solenoid plunger not only trips the spring driven opening mechanism, but also directly drives the opening of the contacts. This "hammer action" provides a very rapid acceleration of the moving contact during the initial contact opening.

In addition to the high current event the circuit breaker must also trip on moderate over-current. This is usually achieved with a bimetallic strip. Increasingly, circuit protection devices provide additional functionality such as residual current or arc fault detection.

1.1 How Circuit Breakers work

The construction of the circuit breaker has been differently and extensively modified to suit various applications. However, the basic components are still distinguishable as shown in Figure 1.1.

The basic technique is to assist interruption by increasing plasma resistance. The arc is drawn between the opening contacts when a short circuit current fault occurs. The current through the conductors of the miniature circuit breaker generates a magnetic field in the arc chamber which acts to force the arc away from the contact region along arc runners and into an arc stack. The arc is then split by a number of parallel plates into a number of series arcs. This results in a high voltage across the circuit as shown in Figure 1.2. The arc voltage works against the system voltage to drive the current towards zero. The result of this is a decrease in the peak current passing through the circuit and reduction in the total admitted to the circuit [1,3,4,5,6,7,8].

Figure 1.1: Circuit breaker structure

The main components of a miniature circuit breaker are the fixed contact, moving contact, arc runners, arc stack and magnetic solenoid trip mechanism. Silver graphite contacts are widely used because they prevent welding after an extended period of resistance heating. When overload occurs, the bimetallic strip heats up and trips the contact over a period of 1-120 seconds. When a short circuit fault occurs, the magnetic solenoid trip mechanism opens the contacts by accelerating a hammer. As the solenoid is energised, the contact opening delay will vary from 0.1-1.0 ms. When a short circuit current occurs, the core of the solenoid is drawn into the coil and accelerated a hammer to trip the contact mechanism. The solenoid is energised by the short circuit current itself.

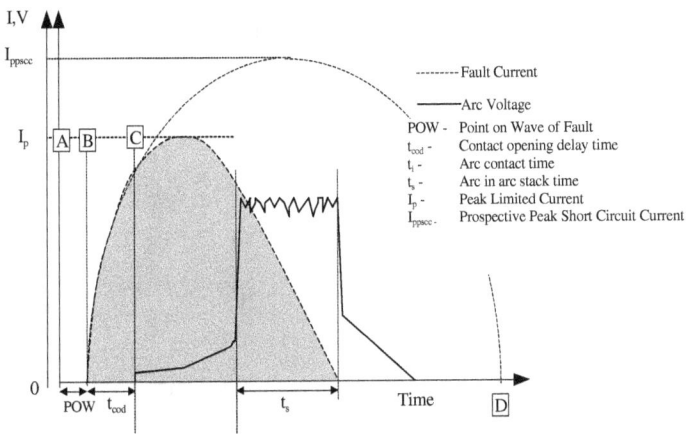

Figure 1.2: The short circuit fault

1.2 Literature Review

The short circuit performance of a circuit breaker is dependent on the rapid opening of the contacts. The arc motion, magnetic fields, gas dynamics of the arc in the arc chamber and high current arc at reduced contact opening velocity has been investigated [1,2,4-9,23]. It has been found that gas flow effects can impede the arc root motion. At low contact opening velocity, the arc root stays longer on the contacts. This commutation delay will generate increased amount of heat in the arc chamber and badly damage the contacts.

It is well established that for reliable operation of the current limiting device the contact opening velocity must be as high as possible with a minimum requirement of 6m/sec. This has resulted in a range of devices where the opening mechanism is triggered by a solenoid designed to be actuated by the short circuit event. In many examples, the solenoid is only actuated by the high current event and is used to provide a hammer action on the moving contact. This results in a very rapid acceleration of the moving contact, resulting in a contact velocity greater than the 6 m/s requirement. In addition to the high current event the circuit breaker must also trip on an over-current, this is usually accommodated with a bi-metallic strip. The complete tripping mechanism therefore has a large number of component parts used in a typical design.

Recent investigations [10-22] have shown that with an optimisation of the arc chamber, resulting from a detailed study of the arc-root phenomena, that the contact velocity can be reduced below the 6 m/sec requirement. This now opens the possibility of refinement of the opening mechanism using smart materials actuation. The closer the match between actuator performance and the demands of the contact system, the simpler the operating mechanism becomes. The number of mechanical parts can be minimised, and the demands on their performance reduced. This will improve reliability, reduce size and power consumption and lower manufacturing costs.

The design of these systems could be simplified by the use of smart materials technology to provide the contact actuation. The closer the match between actuator performance and the demands of the contact system, the simpler the operating mechanism becomes. The number of mechanical parts can be minimized, and the demands on their performance reduced. This will improve reliability, reduce size, power consumption and manufacturing costs. As well as benefits in the design of conventional circuit breakers, the ability to make solid state contact opening mechanisms would allow substantial miniaturization of these devices resulting in a completely new type of device.

The relationship between the arc mobility and the contact opening velocity is complex and depends on the magnetic environment in the arc chamber and the flow of the arc chamber gases. These effects have been investigated previously at reduced contact opening velocity [12-13]. It was found that adverse gas flow effects can impede the arc root motion, and that optimization of the arc chamber design could permit the use of lower contact opening velocities. Recent investigations [10] have shown that detailed study of the arc-root phenomena can yield improvements in arc chamber performance thus permitting operation at reduced contact opening velocity.

1.3 Piezoelectric in Circuit Breakers

This now opens the possibility of refinement of the mechanism using smart materials actuation. Piezoelectric ceramic actuators have been used as the trip element in commercial circuit breakers [5] to provide low power electronic actuation. In this book, the operating characteristics of a piezoelectric actuator relevant to direct contact opening are presented. A directly acting contact opening system was developed. The performance of the contact system under short circuit conditions is analyzed using the flexible test system and arc imaging system [10]. Results are compared with reference data using a constant opening velocity contact system. Successful current limiting performance is demonstrated using the piezoelectrically actuated contact system.

Piezo actuators have used in as the trip element in commercial circuit breakers [24]. Ultimately it may be possible to design a completely solid state contact system.

In this book, the concept of piezo-ceramic actuation in current limiting circuit breakers is presented and discussed. A contact actuation system has been developed and is presented here. The performance of the contact system under short circuit conditions is analysed using the flexible test system and arc imaging system.

CHAPTER 2

PIEZOELECTRIC

2.1 Literature Review

The properties of piezoceramics are highly temperature dependent. Some researches are available on the temperature variation of the low field piezoelectric properties. These are few applications where the piezoelectric are driven at high field strengths [30]. In such circumstances the non-linear effects predominate.

The quantities of non-linear are usually undocumented and data on the temperature variation of these properties [25]. There is some research work available on the effects of temperature on non-linear properties relating to a.c. devices such as sonar and ultrasound [29]. The designer of low frequency systems must accommodate the variations in the static component of the polarisation of the material.

2.2 Hysteresis and Non-Linearity

Hysteresis is a nonlinear phenomenon that can be found in diverse disciplines such as smart materials [54], ferromagnetism and superconductivity [55]. The hysteresis refers to a static memory effect [56], rate independent hysteresis is characterized by (output vs. input) hysteresis loops. Practical hysteresis loops often demonstrate ratedependent behavior due to intimate coupling between hysteresis and traditional dynamics [53].

One of the major obstacles in the application of smart actuators is their hysteretic behaviour and some degree of hysteresis. It has been observed by [53], the hysteresis modelling and compensation methods used for smart actuator devices and compensation is required to provide precision positioning systems.

A typical loop response for a piezoceramic actuator driven by a voltage source is shown in [25]. It was shown the position is multivalued with respect to the input voltage and also it is the response is non-linear.

It is a significant difference between the charging and discharging parts. The repeatability of the loop is good to within a few percent. This fact can be used to achieve highly repeatable open loop operation where the actuation is synchronised to operate on one part of the loop.

The major obstacles in the application of smart actuators are their hysteretic behaviour. If the hysteresis effect is not incorporated into the control system it will act as an unmodelled phase lag whose presence will cause instability in a closed-loop system if sufficient phase margin is not provided in the control design. Integration of actuators with these dynamics into mechatronic systems requires some degree of hysteresis compensation, and also in some cases creep compensation, to provide precision positioning [54].

The degree and severity of hystereis effects in smart actuators can often be reduced by constraining input levels, or incorporating local control feedback. To improve the control of the hysteresis and the linearity can do by controlling the polarisation (charge) rather than the voltage [26]. The system controls the charge on the actuator to achieve highly linear open loop control.

An alternative strategy for controlling hysteresis is to apply a decaying oscillation to a target DC voltage level. The displacement of the actuator converges to a point removing hysteresis from the system. The hysteresis loop converges to the anhysteric line when applied over a range of dc voltages. The non-linearity can be removed by software compensation [25]. In present scanning tunneling microscopic

(STM) designs, PID or robust control laws are used to reduce hysteresis effects [57, 58].

2.3 Re-polarisation

The movement is reversed when a field direction is opposite to a polarisation direction. When the reverse field increases, until it reaches at the point where the polarisation direction changes sign. Thus, subjected to a large amplitude a.c. signal, the characteristic "butterfly" curve is presented as shown in [25].

When the reversal happens, it is normally undesirable since it causes difficulty in control the systems. In addition, the material set up during uncontrolled reversal could lead to degradation in performance.

It has a relationship between temperature and repolarisation voltage. The field strength required to reverse the polarisation is temperature dependent. The re-polarisation voltage continues to decrease towards 0V at the Curie temperature.

To prevent re-polarisation, the system monitors reverse voltages do not exceed a safe level. This enables to deliver maximum performance across the full temperature range without risk of re-polarisation. Future developments will incorporate a temperature sensor to permit compensation for thermal effects.

2.4 Displacement

The deflection of a beam shaped monomorph bending actuator, the d31 effect, which shortens the lateral dimension of the active layer if an electric field is applied in the d33 direction of the polarization. As the active layer is bonded to a passive layer, this causes bending of the device. More efficient designs, known as bimorph, trimorph and multimorph bending actuators, can create bidirectional deflection. By making use of two or more layered piezoelectric structures, similar to piezoelectric multilayer stack actuators, the operating voltage of the multimorph benders is significantly reduced by the small electrode distance [52].

The movement across the temperature range with the reverse charge system is more than 3 times that for a unipolar drive [31]. Re-poling of the ceramic is successfully avoided. This is all accomplished without the use of a temperature sensor or intermediate temperature values. [32]

In recent years there have been some significant advances in extending the concepts of differential flatness and passivitybased control to the infinite-dimension. For applications with large displacements, range of input voltage is available. However, at high electric field strengths, piezoelectric material shows significant hysteretic.

Piezoelectric Trimorph-bending actuators consist of a substrate of metal or carbon fibre and two metalised piezoceramic films [59]. They found that the small structural damping the step-response of the uncontrolled bending actuator has a large overshoot and a large settling time.

2.5 Force deflection

Increasing the deflection by a factor of at least 10 is desirable for the majority of applications, even if a loss of force must be accepted. This remarkable performance is achieved by bending type operation by means of a special mounting. The high tensile stresses involved in bending mode operation, the reliability of this actuator has to be investigated very carefully. To increase acting force the actuator elements should be arranged in the plane of the piezoelectric layer. Higher deflection values can be achieved by stacking of the bending elements [60].

It shows the force deflection characteristic at room temperature. This is a typical characteristic of computerised load testing machine (Lloyd Instruments). The stroke is defined as the difference between the on and off positions. The actuator exhibits a linear force displacement profile with, in this case, a stroke of 1.21 mm, blocked force of 0.18 N and a stiffness of 151 N/m [27].

This shows the relationship between the voltage and the charge loops at room temperature and also the characteristic non-linear hysteresis loop when driven by an oscillating (sine wave) voltage.

The profiles of the charge and the position are quite linear with almost no hysteresis. It has been used previously to improve the linearity of the piezo response [26, 28].

2.6 Temperature Variation

The relationship between temperature variation and charge cycles (sine wave) shows that the depolarisation of the ceramic consists of a flow of charge around the circuit. This curve is therefore related to the thermally generated charge produced by the pyroelectric effect.

With temperature variation of the slopes of charge cycle, this is the amount of movement produced by a given amount of charge reduces as the temperature increases.

This is quite linear and is described by: [26]

$$\frac{dz}{dQ} = AT + B$$

Where: $A = -0.54 \; \mu m \; \mu C^{-1} \; K^{-1}$
           $B = 113 \mu m \; \mu C^{-1}$

It is known the pyroelectric charge is related fairly linearly to the temperature by the pyroelectric coefficient.

The slope of the pyroelectric line is a charge coefficient for the actuator dQ/dT. These linear charge relationships can be used to approximately evaluate the non-linear depolarisation using simplified relationships between the quantities:

$$\frac{dz}{dT} = \frac{dz}{dQ}\frac{dQ}{dT} = (AT + B)\frac{dQ}{dT}$$

$$z = \int \frac{dz}{dT}\,dT = \left(\frac{1}{2}AT^2 + BT\right)\left(\frac{dQ}{dT}\right)$$

The depolarisation curve shows the correct order of magnitude and the fit of the slope improves at lower temperatures. This could be due to difference between the thermal coefficients and the substrate which could also cause the low temperature upturn in the depolarised line [27]. If the data are compensated for this then a good fit is obtained up to the point of actuator closure.

The design of a reliable piezo operated mechanism for circuit breakers must allow for variations in the actuator on and off. The thermal variations in these positions relative to the mechanism are a key factor in the design of circuit breakers.

The understanding of the thermal effects described in this book permit optimum placement of the actuator relative to the mechanism. These lead to the operating window, reducing and relaxing the allowable variation in parts and improving the reliability of the switching devices.

Piezoelectric actuators are usually glued onto the surface [44, 45]. The stack actuators are mounted into the force flux by interrupting the passive metal structure with mechanical fixation for the stack actuators [46, 47].

The potential of active components can be fully exploited if the active components are directly integrated into the metal parts [48]. For manufacturing process, it was shown that integration of piezoceramic plates and modules into aluminium parts during a common die casting can be done [49, 50].

There is some research [51] on how far piezoceramic stack actuators survive the high thermal loads of the die-casting process and if their long term reliability is affected. It found that the degree of the ferroelectric depolarisation is correlated to the duration time in the melt as well as to the temperature and the heat transfer at the surface. The results show that the effect of heat introduction into piezoelectric actuators due to the die-casting process does not necessarily lead to a severe degradation.

2.7 Harsh Environments

There is some study about leakage current during humidity tests using piezoceramics as DC actuators in harsh environments [38]. Under humid conditions the piezoelectrics underwent some permanent changes properties resulting in increased leakage currents.

The PE loops measured in the humid environment show a much greater hysteresis thannormal ambient. This is because of the large leakage current contribution. Thus, in the hot and humid conditions, there is a steady increase in leakage current. This increases the power consumption. There is an increase in conductivity of actuators in humid conditions under DC bias.

Garbuio and Rouchon [61] have studied the piezoelectric thrust bearing for severe environments. They show the modeling of the physical principle results to an actuator. It is possible through active membrane to control the resistive torque ensuring blocking, the braking and the disengaging of any mechanism by friction. These forces can be significant and in conditions of dry friction what makes it possible to consider such devices under severe operating conditions.

This actuator can then be used under severe environment because of completely isolated from outside. From the geometry of the actuator considered, the mechanical decoupling of the vibrating end part of the actuator with respect to the frame also allows to exploit the mechanical vibration and its effects through an active surface.

2.8 Micro Actuators

The electrostatic and piezoelectric interaction requires large driving voltages, while the magnetic actuation potentially provides both large force and large displacement in an energy-efficient manner [39]. Wagner et al. [40] have proposed manually attached precision-machined permanent magnet pieces on suspended plates with integrated in-plane coils to generate an external magnetic field.

A toroid and a quadrupole coil design for microactuators have been studied by [41] in both numerical and experiments. The interaction between the soft ferromagnetic core and the permanent magnet might be unfavourable for an actuator due to the decreased controllability.

If the actuator is used as a micro-resonator this behaviour does not account when placing the permanent magnet at the point of maximum force. The variation of the small force induced by the coils might lead to significant membrane amplitudes by using its first resonance frequency.

A linear hybrid micro actuator with a hybrid step motor drive scheme is developed by [62], the vertical force of the linear micro actuator greater than the driving force. For linear micro actuators, variable reluctance and hybrid step motor drive schemes are particularly well suited [63]. The key technology challenges were the development of a dual layer spiral coil, a process to deposition of thick permanent magnets with a high energy product. This is required as permanent magnet material [64] and an assembly technology to allow a micron range gap between stator and traveler.

2.9 Honeycomb Piezoelectric Actuators

Multilayer piezoelectric actuators are often used for an advantageous in their high response and high power [42]. Low-cost honeycomb piezoelectric actuators comprise a piezoelectric quadratic honeycomb structure and electrodes disposed on inner wall surfaces of the cells.

The process of mass-production was shown in [43] by extrusion as catalysis carrier. By applying a voltage between the electrodes adjacent through the partition walls, this actuator can be contracted in an axial direction. The advantages are high durability and lightweight.

The deflection of a beam shaped monomorph bending actuator, the d31 effect, which shortens the lateral dimension of the active layer if an electric field is applied in the d33 direction of the polarization.

As the active layer is bonded to a passive layer, this causes bending of the device. More efficient designs, known as bimorph, trimorph and multimorph bending actuators, can create bidirectional deflection. By making use of two or more layered piezoelectric structures, similar to piezoelectric multilayer stack actuators, the operating voltage of the multimorph benders is significantly reduced by the small electrode distance.

CHAPTER 3

PIEZOELECTRIC ACTUATORS

The development of piezoelectric actuator in circuit breakers is a new technology for electrical protection systems. A range of alternatives have been investigated such as the redesign of the normal trip mechanism to permit smaller magnets.

3.1 Planar Bimorph Actuator

The actuator used for the present investigation in this book was a planar bimorph actuator shown in Figure 3.1.

D31 benders give better motion. To increase the motion and force, a rectangular actuator form to increase the anisotropy of the beam and a cantilever system. It gains a motion in order to interact with a mechanism. This is because a cantilever has many times the deflection of a simply supported beam of the same section and length [24, 25].

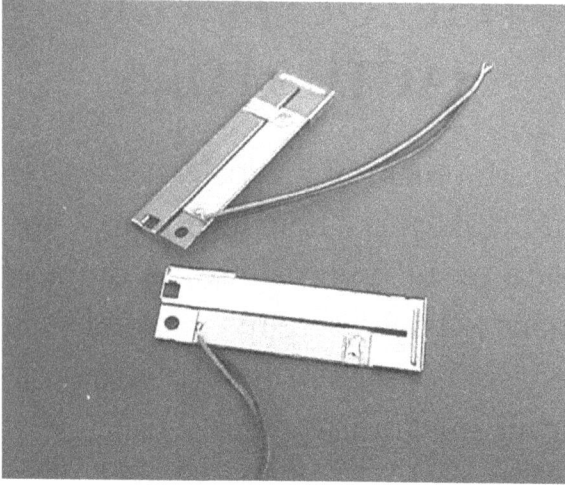

Figure 3.1: Planar bimorph actuator

The actuator employs a novel geometry, which produces very large movement from a simple and compact structure [24].

3.2 Planar Bimorph Operation

The actuator comprises two arms formed from a single metal substrate and two piezoelectric ceramic plates. The operating principle is shown in Figure 3.2.

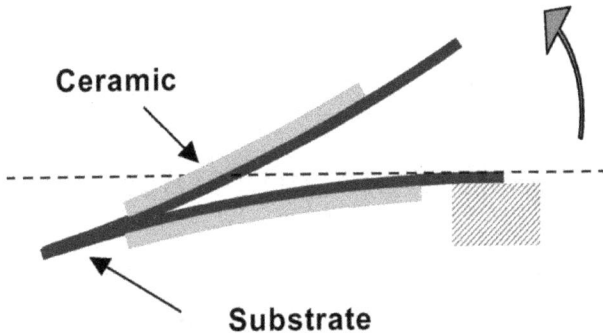

Figure 3.2: Schematic diagram of the operation of the planar bimorph actuator [16, 18]

The planar disposition of the two ceramic plates means that the high fields can be applied without danger of repolarising the ceramic. This is different from a conventional bimorph structure.

The first arm of the actuator produces a downward movement which provides an angle for the movement amplification of the second arm. The total movement is slightly less than could be obtained from a single beam of twice the length.

However, the stiffness of a single beam has cubic relationship to the length. So that only very low forces could be obtained this way.

3.3 Planar Bimorph Structure

A key advantage of the planar bimorph structure is that it forms an effective pivot point approximately half way along the length of the actuator as shown in Figure 3.3.

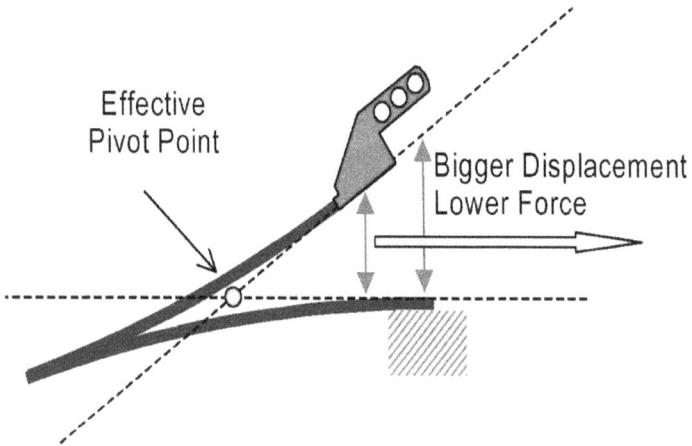

Figure 3.3: Effective pivot in planar bimorph actuator [16, 18]

Therefore, a high displacement can be obtained with useful force output. This results in a much more efficient actuator than a conventional single beam design.

The piezoelectric ceramic is a commercial grade lead zirconate titanate. It is selected for its high coefficient 330 pm V to achieve a large displacement in a bending actuator. The formed and compacted ceramic is sintered and then machined to size. The details of electro-ceramic processing have shown in [33]. This forms a dense grain structure with typical grain sizes of 15- m diameter, as shown in the surface profile of Figure 3.4. A metallic electrode is then applied to the ceramic. The ceramic plates are bonded to the metal substrate and electrical connections made to complete the assembled actuator.

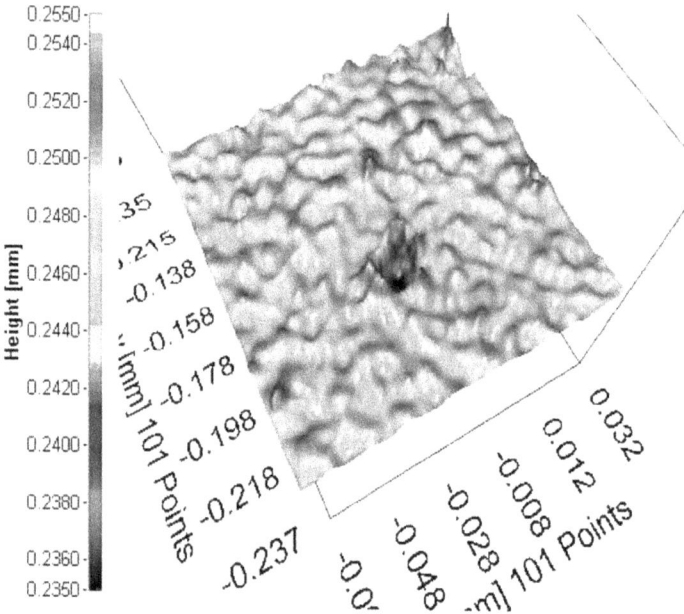

Figure 3.4: Surface structure of piezoelectric ceramic (Image courtesy Taicaan Technologies)

3.4 Piezoelectric in Circuit Breakers

When applied in a circuit breaker mechanism the actuator behaviour can be approximated by a simple mass—spring system [16, 18]. The spring rate is independent of the applied voltage, but the unloaded position varies with voltage.

However, the complex thermal behaviour of the piezoceramic materials can often severely limit the performance of the actuation system. This leads to increased cost, complexity and reliability issues.

Not only do the basic piezoelectric parameters vary with temperature, but also it is dominant. The effect is the rapid increase in the hysteresis as the temperature reduces. This is especially true for the very active soft ceramics used to provide maximum performance from the actuator.

Moreover, the coercive field vary with temperature. The relationship between position and applied charge or voltage is also temperature dependent.

It is possible to implement a control system that applies a temperature variable reverse field and charge rate using a low cost microcontroller. This system gives substantially improved performance at low temperatures. In fact the actuator force and deflection increases as the temperature decreases.

This improves tolerance to variations in position of both the actuator and the mechanism being actuated. This can lead to cost benefits in the deployed mechanism through the use of lower precision manufacturing and assembly techniques.

CHAPTER 4

INSTRUMENTATION

4.1 Introduction

A schematic diagram of the arrangement apparatus and associated instrumentation is shown in Figure 4.1. Short circuit tests were carried out using a Flexible Test Apparatus (FTA) designed to simulate the current limiting operation of a miniature circuit breaker. The high speed Arc Imaging system (AIS) used to record the motion of the arc. The test system has previously been described in [10-23].

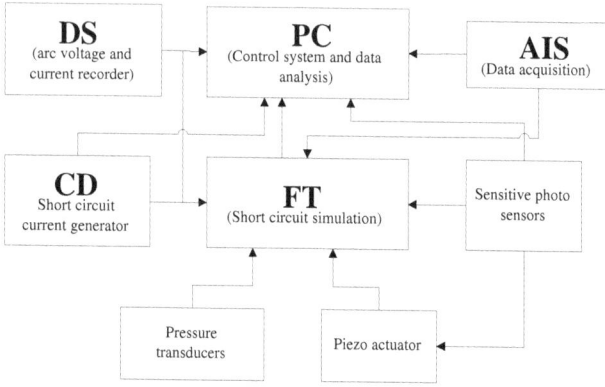

Figure 4.1: Schematic diagram showing the arrangement of the apparatus and associated instrumentation

Computational manipulation of optical data permits the identification of parameters describing the motion of both cathode and anode arc roots on the fixed and moving contacts. These instrument techniques were used to study the influence of arc chamber material, contact material, and contact opening speed on the arc root mobility during contact opening. The arc chamber parameters were varied using the Flexible Test Apparatus (FTA) described previously [10,11,12]. The arc root mobility in the contact region has been investigated individually on the cathode and anode arc root movement.

4.2 Flexible Test Apparatus (FTA)

The Flexible Test Apparatus (FTA) was used to simulate the operation of a miniature circuit breaker when a short circuit current occurs. The highlights of the FTA are the independent configuration of the contact mechanism, repeatable contact action, variable contact material, variable arc chamber geometry, variable arc chamber and variable arc chamber vent configuration.

The overall details of the FTA are shown in Figure 4.2 to 4.6. The FTA is bolted to an earthed aluminium plate. The structure of FTA is similar to commercial MCBs. The main components consist of solenoid, arc chamber, moving contact, fixed contact, arc runner, arc chamber vent and arc stack. An array optical fibre is placed on top of the arc chamber. The arc chambers are machined from Tufnol material. Arc chamber sidewalls are made from Macor machineable ceramic.

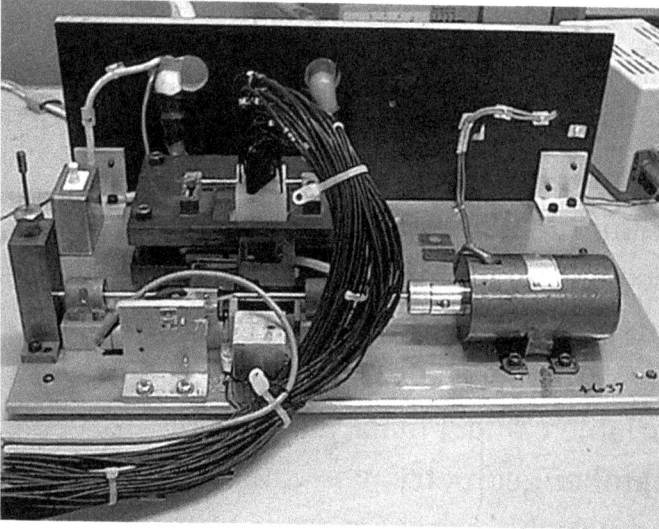

Figure 4.2: Flexible Test Apparatus (FTA)

The top plate is a transparent clear view to allow the arc motion to be recorded by the AIS. The bottom plate is changeable to study the effect of arc chamber material. The moving contact mechanism is opened by the impact of a hammer connected to the solenoid. When the hammer passes the optical sensor, the trigger signal is sent to count down at the control computer to start the short circuit current fault and data acquisition.

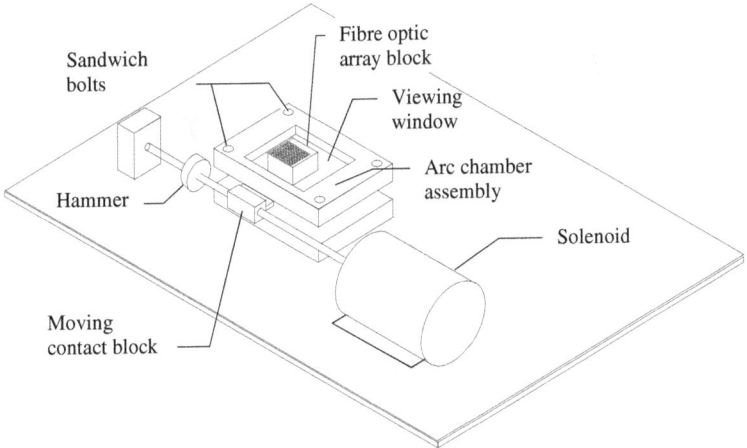

Figure 4.3: Structure of the Flexible Test Apparatus (FTA)

The dimension of the arc chamber and position of the fibre optic is shown in Figure 4.3. Chamber length l = 45 mm, Chamber width = 29mm, Chamber depth d = 6mm.

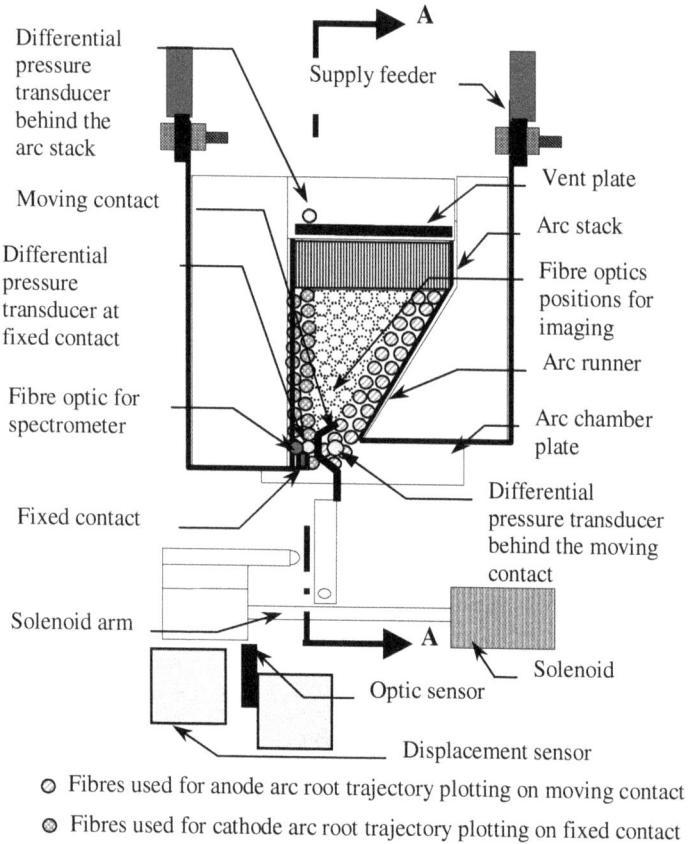

Differential pressure transducer behind the arc stack

Supply feeder

A

Moving contact

Vent plate

Arc stack

Differential pressure transducer at fixed contact

Fibre optics positions for imaging

Arc runner

Fibre optic for spectrometer

Arc chamber plate

Fixed contact

Differential pressure transducer behind the moving contact

Solenoid arm

A

Optic sensor

Solenoid

Displacement sensor

○ Fibres used for anode arc root trajectory plotting on moving contact

◎ Fibres used for cathode arc root trajectory plotting on fixed contact

Figure 4.4: Flexible Test Apparatus component

A cross section of the arc chamber in the Flexible Test Apparatus (FTA) and the arc chamber are shown in Figure 4.4 to 4.6.

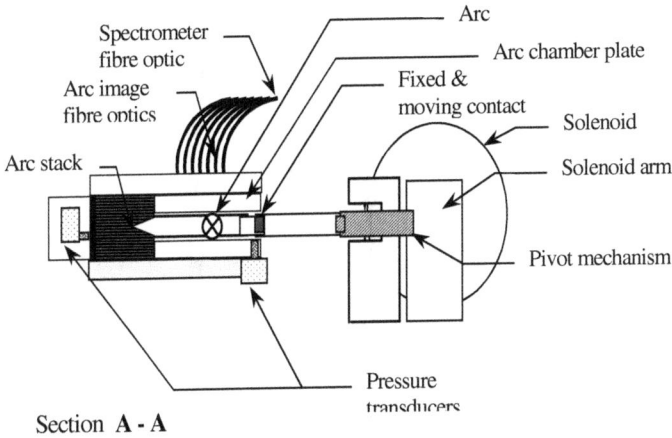

Section **A - A**

Figure 4.5: Cross section of the Flexible Test Apparatus (FTA)

Commercial devices are variable. Many of them use ceramic side plates in the arc chamber. The ceramic is a good insulator, but also acts as a heat sink. Ceramics are used in MCBs , but may be more prevalent to larger circuit breakers such as moulded case breakers. The ceramic arc chamber in the Flexible Test Apparatus (FTA) would be quite similar to this.

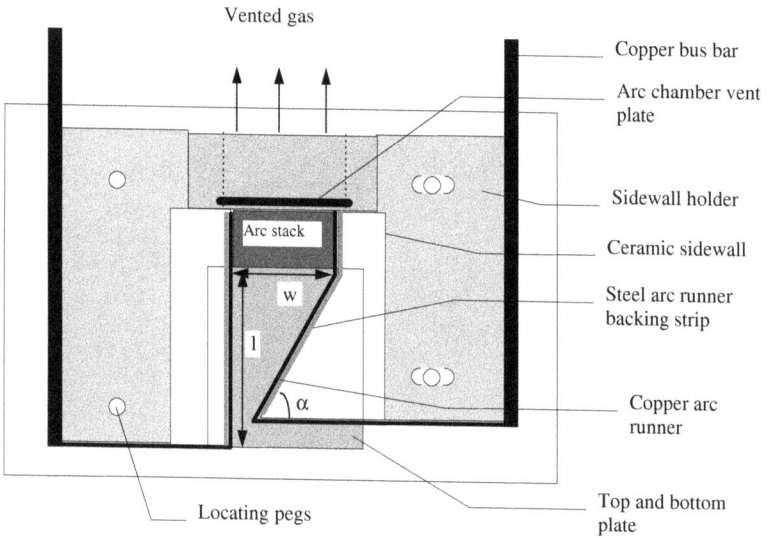

Figure 4.6: Arc chamber geometry in the Flexible Test Apparatus (FTA)

Some devices use plastic parts for the side walls which would have different thermal properties. The effects of different materials combinations are part of the research in this book. This is a substantial body of research into different materials and their effect on the arc control. Ina way, this whole area of research needs to be re-applied to the optimisation of a low contact velocity device.

4.3 High Speed Arc Imaging System (AIS)

The Arc imaging System (AIS) is employed to record the optical data from the arc chamber in the Flexible Test Apparatus (FTA). Each optical fibre has a defined position relative to the interior components of the miniature circuit breaker. Polymer fibre with 1 mm. core diameter surrounded by a 0.5 mm sheath was used. The attenuation of polymer is 200 dB/km at 665 nm [23].

A dimensional array of fibre optics is placed over the arc chamber of the FTA, as shown in Figures 3.4. Each fibre optic records the light level at a specific location within the arc chamber. The arrangement of the fibres within the fibre optic array was customised for each arc chamber geometry used in the investigations. The location of the fibres used for each arc chamber is shown in Figure 3.4 and Figure 4.2, Chapter 4. The fibres are hexagonally close packed with centres 3 mm apart. Each optic fibre in the array is recessed in the fibre optic array in a round hole as shown in figure 3.6.

The radius of view at the centre of the arc chamber depth is calculated by

$$r = t \ (0.5 + d/a),$$

where: t is the fibre diameter (1mm),

d is the depth to the viewing plane (15 mm)

a is the depth of the fibre recess (25mm). This gives a viewing radius of 1.1 mm for each fibre optic [2].

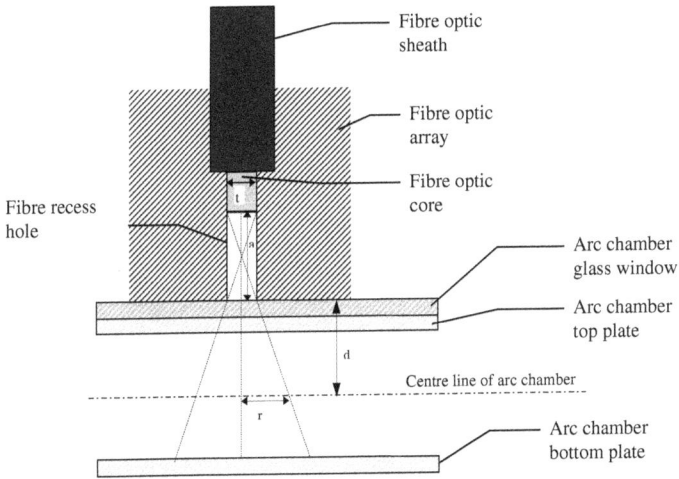

Figure 4.7: Fibre optics acting as a pinhole camera

The light from each fibre is converted to an electrical signal by the optical circuit as shown in Figure 4.7. A photodiode is used to convert the light transmitted through the optical fibre into an electronic signal. Then the analogue signals are multiplexed into a group of eight and then converted to digital format by A/D converter as shown in Figure 4.8.

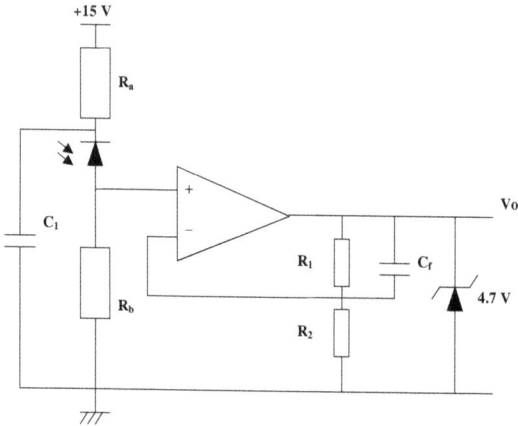

Figure 4.8: Photo-detector and amplifier circuit

Data is retained in the random access memory (RAM) of the AIS. A digital I/O card in the personal computer is then used to transfer the data [10-16]. The schematic diagram of the AIS is shown in Figure 4.9.

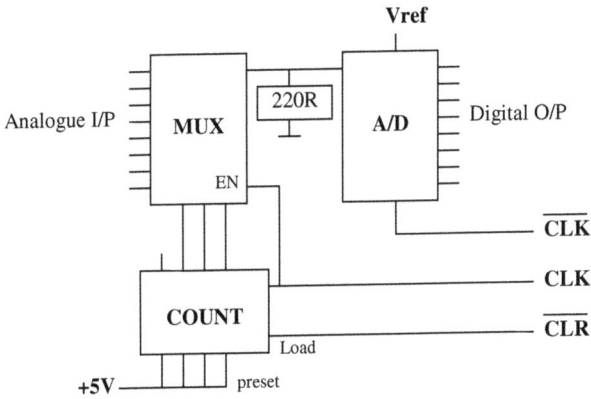

Figure 4.9: Analogue multiplexer and A/D converter

The Schematic diagram of data recording in the Arc Imaging System is shown in Figure 4.10. The AIS samples light levels from each fibre optic at a rate of a million samples per second, and it is capable to record the arc movement for 8 ms. Each complete data acquisition card can handle 15 fibre optics cables. Each card features 15 photo-transistors, two 8 way multiplexers, two 6 bit 8 MHz Flash A-D converters, 32K of Ram and also the control & timing circuit [10, 66].

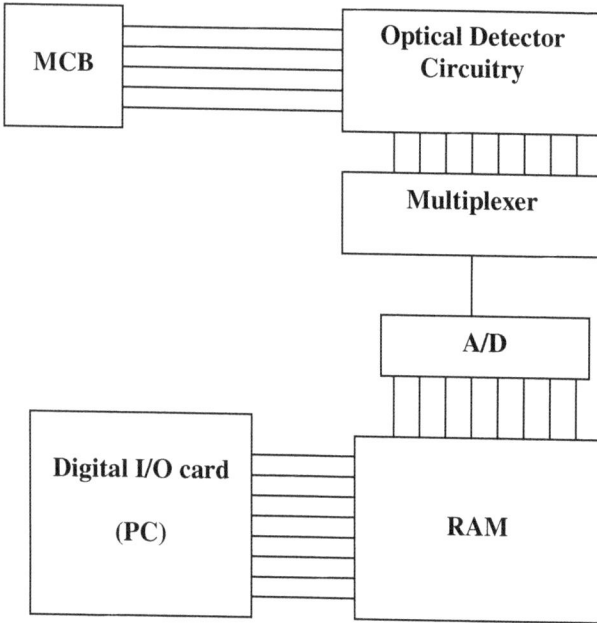

Figure 4.10: Schematic diagram of data recording in the Arc Imaging System

4.4 Pressure Instruments

Two pressure transducers are used to monitor the gas pressure in the arc chamber. The SX series of piezo resistive pressure sensor functions as a wheatstone bridge on a silicon chip with a response times of 0.1 ms. The pressure transducer gives a voltage output directly proportional to the applied pressure with sensitivity 0.75 mV/Psi. Long term stability is 0.1% and repeatability is 0.5% (Maximum difference in output at any pressure with the operating pressure range and temperature within O °C to +70 °C). Resistors are ion implanted into the silicon. The cavity etched on the reverse to create a thin silicon diaphragm.

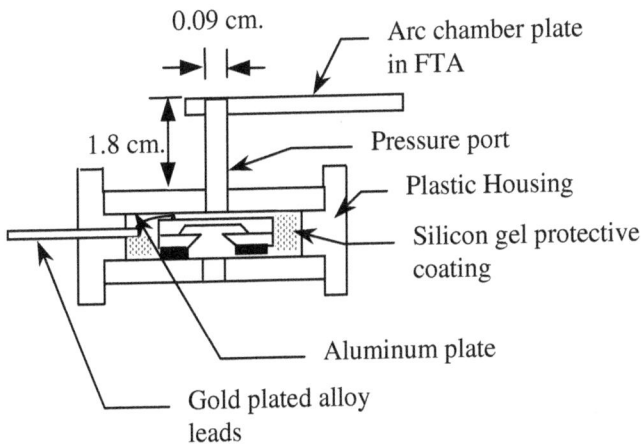

Figure 4.11: Installation and connection of pressure transducer

The pressure transducers were installed in the fixed contact region, in the gap behind the moving contact, and behind the arc stack. The locations of pressure transducers as installed in the FTA are shown in Figure 4.4. The pressure transducer connection and pressure transducer circuits are shown in figure 4.10 and 4.11.

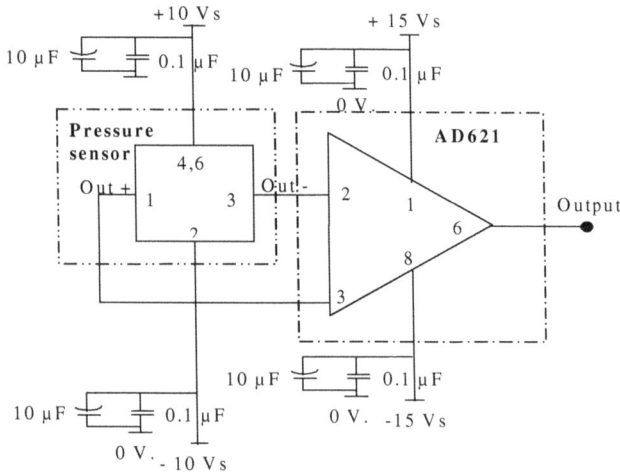

Figure 4.12: Pressure measurement circuit

The instrument amplifier AD621 was used to amplify the signal. The pressure transducer instrument was calibrated against a gauge pressure standard.

Differential pressure measured relative to atmospheric.

4.4.1 Accuracy of pressure measurement

The accuracy of the pressure measurement was tested as details as follows:

Vibration effect: This is concerned about the vibration on the pressure transducer transmitted through the Flexible Test Apparatus (FTA). This was shown to be minimal.

Environment effect: This is concerned on the effects of the environment (air inside the tube with 1.8 cm. Length). This tube is connected from a pressure transducer to the arc chamber bottom plate in such a way to minimise environmental effects.

4.4.2 Electrical Noise

Power supplies: Each pressure transducer is connected to a difference power supply and separated amplifier circuit.

Cables: Three power cables from +Vs, GND and -Vs are twisted together to protect cross talk and noise.

Length of cable: the length of cables is limited as short as possible and close to the pressure transducer circuit.

4.4.3 Ground system

All of ground connectors for all experiments were connect together to the earth aluminium plate and aluminium boxes of the amplifier circuit.

4.5 Piezoelectric Instrumentation

The Piezo-ceramic used as the actuator in this paper is Lead Zirconate Titanate (PZT). The actuator is designed in a U-shape, as

shown in Figure 4.13. The material is
contructed in a sandwich fashion with layers
of PZT material. This structure and the
assembly in the U-shape arrangement
provide maximum forward deflection of 4
mm and maximum reverse deflection 1.5
mm.

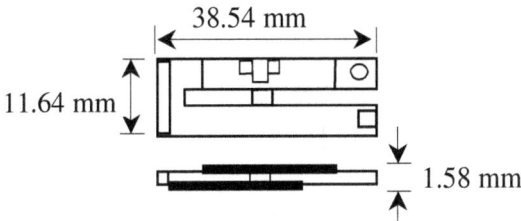

Figure 4.13: Piezoelectric dimensions

Figure 4.14: Displacement laser and instrumentation

The full specification of the material used is provided in Table 4.1.

Electrical performance	
Nominal voltage	325 Vdc
Max. forward voltage	500Vdc
Max. reverse voltage	75 Vdc
Physical performance	
Nominal deflection per volt	10^{-6} m.
Max. forward deflection	4 mm.
Max. reverse deflection	1.5 mm.
Time to deflect (min)	1.25 ms
Blocking force	0.2 N
Beam release force (max)	2.5 N
Physical data	
Ceramic	Lead, Titanium, trace
Substrate	Alloy 42 (Ni,Fe,Si,Mn)

Table 4.1 Characteristics of piezoelectric

A Laser Triangulation Sensor was used to measure the deflection of the piezoelectric. The sensor has a 0.01μm resolution. The power supply for piezoelectric was used dual step up transformers transform from 230Vac to 360 Vdc via variable resistor. More than 200 data and 12 sets were recorded to study the relationship between supply voltage and deflection of the piezoelectric.

CHAPTER 5

EXPERIMENTAL SETUP

5.1 Contact speed setup

The opening contact mechanism was modified to reduced contact speed from 10 m/s down to 1 m/s. There are two important parts in this system. The first part is the instrument to measure the speed of contact and the second part is the instrument to control the speed. The instrumentation used to measure the contact velocity is the displacement sensor which is connected to the solenoid. The method used to control the contact velocity is to control the supply voltage to the motor of the solenoid.

5.1.1 Displacement sensors

The pivot mechanism moves in a curve with a very high speed, it is very difficult to measure speed directly from this system. It is not very difficult to measure the velocity of the solenoid movement and calculate the speed of the pivot mechanism and contact. The displacement transducer uses a filter pattern together with intensity sensitive photo sensors. Details of the displacement transducer control circuit and differential op-amp circuits are shown in Figure 5.1 –5.2.

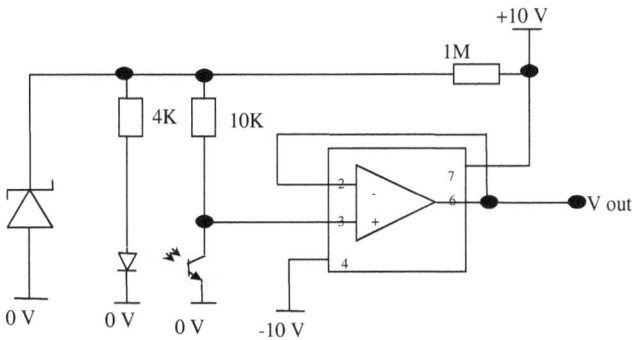

Figure 5.1: Control diagram of displacement transducer

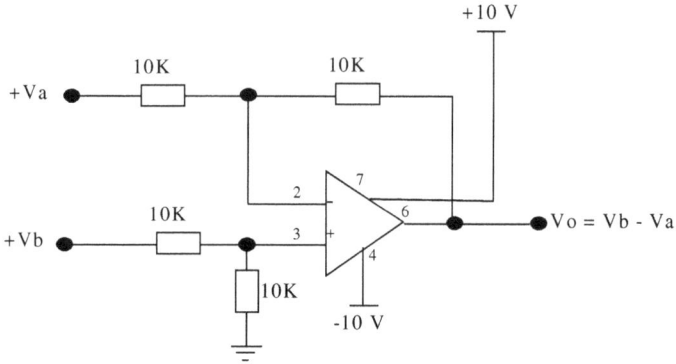

Figure 5.2: Detail of the differential circuit for output signal of displacement transducer

5.1.2 Contact speed control

The velocity of the solenoid movement can be controlled by reduced power supply voltage. The voltage across the armature and the strength of the magnetic field is the main factor which affects the speed of the solenoid [68]. This can be controlled by many methods i.e. shunt field, armature resistance, variable voltage and multi-voltage. In this research work, the armature resistance method is used to control speed. The wiring diagram to control contact opening velocity by inserting a variable resistor in series with a supply solenoid is shown in Figure 5.3 – 5.4 to reduce the speed of solenoid. The supply voltage of the solenoid is decreased due to the voltage drop across the series resistance. The contact velocity can be regulated between 10 m/s and 1 m/s with average tolerance ±0.02 m/s.

The contact opening velocity is measured as the displacement of the solenoid arm while moving passes the optic sensor. For repeatability of the contact opening velocity from 10 m/s to 1 m/s, more than 15-20 experiments have been tested. The accuracy of the contact opening velocity of 10 m/s is approximately ± 0.01 m/s and ±0.02 m/s for contact opening velocity at 5.5 m/s, 4.0 m/s and 1 m/s.

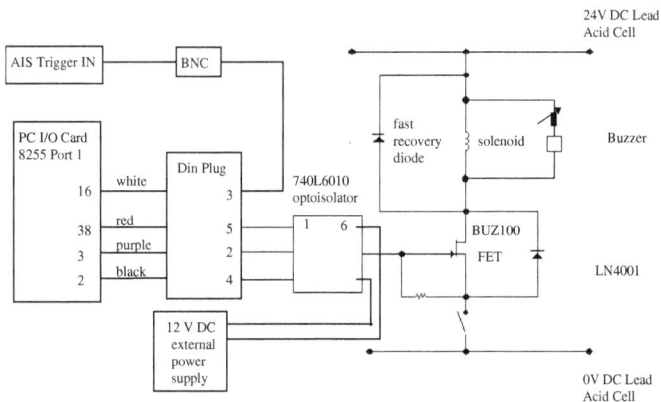

Figure 5.3: Existing solenoid control circuit for FTA [2]

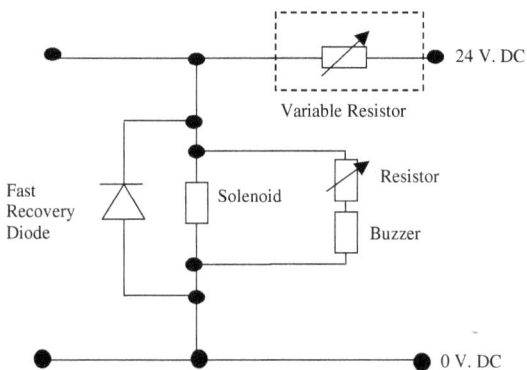

Figure 5.4: Modify solenoid control velocity

5.2 Experimental Parameters

5.2.1 Experimental variable parameters

There are seven variable parameters used to investigate the effects on the mobility of the arc root from the contact region. They are contact material, arc chamber vent, peak short circuit current level, contact velocity, arc chamber material, the gap behind the moving contact and supply polarity. The details of the variable parameters to be investigated are shown in Table 5.1 and shown in Figure 5.5.

Contact opening velocity	10 m/s, 5.5 m/s, 4 m/s, 1 m/s
Supply polarity	- Anode on the moving contact and cathode on the fixed contact - Anode on the fixed contact

	and cathode on the moving contact
Peak short circuit current level	2000 A, 500 A, 1400 A.
Contact material	Ag/C (95/5) step on the moving and fixed contact Cu punch on the fixed contact Ag/C (95/5) flat on the fixed contact
Arc chamber venting	Opened, Choked, Closed
Arc chamber material	Ceramic, Plastic polycarbonate
Gap behind the moving contact	Opened, Closed

Table 5.1: Experimental materials

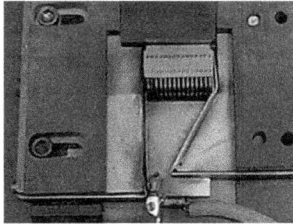

Component parts in the FTA

Ag/C flat on the fixed contact

Cu punch on the fixed contact

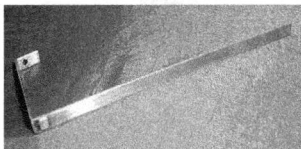

Ag/C step o the fixed contact

Moving contact arc runner (typical)

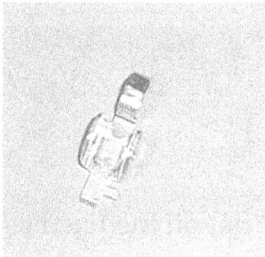

Ag/C step on moving contact

Moving contact (typical)

Polycarbonate and ceramic arc chamber

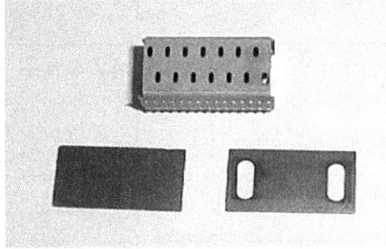

Vent: Opened, closed and choked

Figure 5.5: Experimental materials

5.2.2 Experimental fixed parameters

The experimental study is carried out on the arc root motion when the arc is drawn from the fixed and moving contacts until the arc reaches the arc stack. The arc displacement, arc current, arc voltage, pressure in the arc chamber and spectra of the arc are presented for each experiment. The behaviour of the arc is then analysed from these plots. Experimental constants for these experiments are defined in Table 5.2.

Experimental Factor	Fixed value
Discharge system Inductance	224 µF
Discharge system Capacitance	47.4 mF
Final contact gap	6 mm.
Contact mechanism	Pivoting mechanism
Moving contact	Silver plated copper
Contact opening delay (t_{cod})	500 µs

Table 5.2: Experimental constants

5.3 Flexible Test Apparatus (FTA) setup

The piezo actuator has then been assembled into the Flexible test apparatus as shown in Figure 5.6. The experimental test chamber is shown with the arrangement of the optical fibre array used for the evaluation of the arc root motion [11, 18].

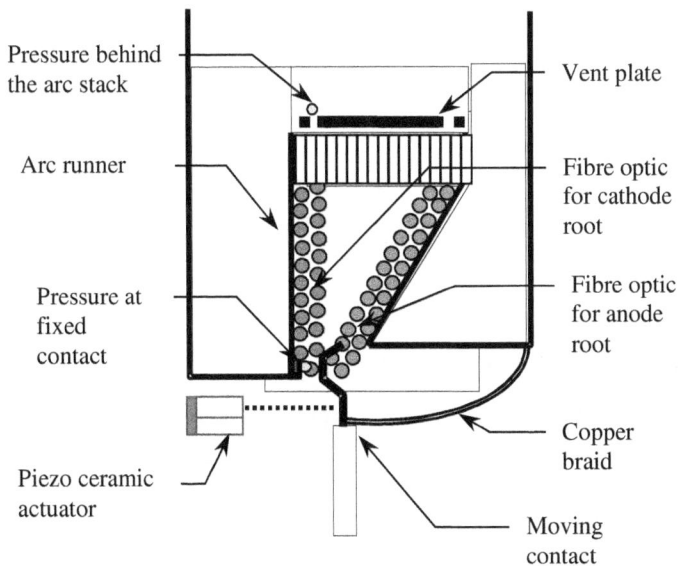

Figure 5.6: Schematic diagram of the Flexible Test Apparatus and piezo ceramic actuator

All surfaces are cleared after the arc chamber wall is removed from the test rig. The new copper arc runner and the steel backing plate are cleaned before they are fixed into the arc chamber. The moving contact is cleaned before it is soldered to connect with the copper braid. After that, the moving contact is fastened into the moving contact block assembly.

The copper braid is fixed with the moving contact runner into the arc chamber wall. The arc stack is put behind the arc chamber. A quartz glass is inserted above the arc chamber until it touches closely to the arc stack.

The top Tufnol plate with the fibre optic array is placed over the top of the arc chamber and fixed with bolts. Now the Flexible Test Apparatus (FTA) is ready to start a new experiment. However, it is important to make sure that the current flow continuously between the fixed contact and the moving contact. The experimental results are presented here as an investigation of the arc root contact time, the average and the standard deviation of the anode and cathode root contact time on the fixed contact and on the moving contact.

After all of materials are prepared in the Flexible Test Apparatus (FTA). The solenoid's arm, which directly connects to the hammer, is manually moved to the maximum displacement. Next, The Digital Storage Oscilloscope (DSO), the Arc Imaging System (AIS), the control computer and the solenoid power supply are switched on.

5.4 Pressure transducer setup

To investigate the influence of the pressure in the arc chamber on the arc root movement, the pressure transducers are installed in three locations inside the arc chamber.

A two layer thermoplastic and metal tube is connected into the pressure transducers. The ends of these tubes are glued together to the base of pressure sensor to protect gas leaking. The pressure sensor is then connected into the pressure measurement circuit, amplifier circuit and power supply. The output probe is connected into the digital storage oscilloscope (DSO). Before using the pressure transducers in the experiments, all of the pressure transducers are tested with a standard gauge to make sure that the error of the pressure measurement is minimal. This also provides the linear relationship between output (voltage) and pressure (bar).

The tube of pressure transducer is connected into a bottom plate of the arc chamber. There are three locations to install the pressure transducers in the arc chamber: in the fixed contact region, behind the moving contact and behind the arc stack. To measure the pressure behind the arc stack, an extra thermal plastic sheet is used to cover the tube and pressure transducer which install behind the vent plate next to the arc stack.

The short circuit arc current is used as a trigger signal for the pressure transducers to capture the pressure signal from the arc chamber. The output signal is recorded in a memory of the digital storage oscilloscope (DSO) as a real time function from the arc ignition until the arc extinguish. Afterward, the program "Flexgav4.base" downs load the data stored on the memory of the Digital Storage Oscilloscope (DSO) into the hard disk of computer (PC) via interface bus for further analysis.

CHAPTER 6

EVALUATION METHODS

The Flexible Test Apparatus (FTA) is used to simulate the operation of a miniature circuit breaker when a short circuit current occurs. The Arc imaging System (AIS) is employed to record the optical data from the arc chamber in the Flexible Test Apparatus (FTA).

Two pressure transducers, the SX series of piezo resistive pressure sensor, are used to monitor the gas pressure in the arc chamber. Flexgav4.bas is the programme to run the test sequences operation. Arcimage.bas is the program to present a series of the arc motion. Rootplot.bas is programme to analyse the experimental results.

The arc chamber geometry and contact opening velocity are modified. The velocity of the solenoid movement is controlled by the voltage across the armature of the stepping motor. To limit the gap behind the moving contact, a piece of Macro ceramic is used to close the gap behind the moving contact.

6.1 Method for evaluating arc root contact time

The optical data from the AIS allows a study of the arc root contact time on the fixed and moving contact individually. The arc position is calculated from the numerical data, the position of the Centre of Intensity (COI) of the arc for each sample period from the light intensity distribution of the whole arc. The products of the light intensity and the position were summed over the whole array, and then divided by the sum of the total light intensity,

$$x = \frac{\sum I_i X_i}{\sum I_i}$$

$$y = \frac{\sum I_i Y_i}{\sum I_i}$$

The result of the calculation gives the X and Y co-ordinate of the average light intensity for each time sample period. The position of the arc is subsequently defined as the position of the centre of intensity of the light at any point in time [2, 10]. The position of the arc root can be determined by the position of the light intensity along the selected fibres along the arc.

The arc voltage and arc root trajectories are shown in Figure 3.21 to 3.22. The arc voltage is shown as the lower trace. The cathode root and anode root on the moving contact and on the fixed contact as shown in the upper trace are studied individually. To allow a full analysis of parameters, the arc root contact time on the fixed or moving contact is defined as follow:

6.1.1 Arc root contact time on the moving contact

The anode and cathode arc root contact time on the moving contact is defined as the time difference between the start of the arc (The point that the arc voltage wave form rises up rapidly) and at that point that the arc root displacement passes a 10 mm as shown in Figure 6.1.

The arc root contact time on the moving contact is shown to start moving away from the contact region at approximately 1550 μs. The contact opened and the arc occurs at the time 468 μs. Therefore, the arc root contact time on the moving contact is approximately 1550-468=1082 μs.

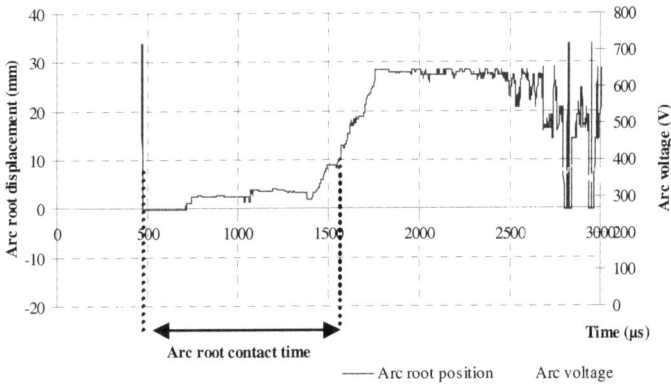

Figure 6.1: Arc root contact time on the moving contact

6.1.2 Arc root contact time on the fixed contact

The anode and cathode arc root contact time on the fixed contact is defined as the time between the start of the arc (The point that the arc voltage wave form rises up rapidly) and the time at that the arc root displacement moved away from the region at 0 mm as shown in Figure 6.2.

The arc root contact time on the fixed contact is shown to start moving away from the contact region at approximately 1100 μs. The arc is started at the time 468 μs. The arc root contact time on the fixed contact is approximately 1100-468=632 μs.

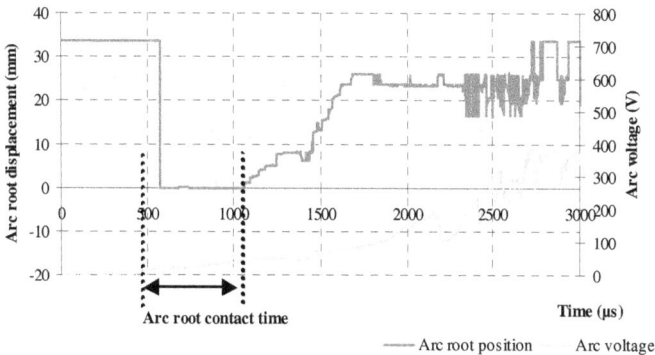

Figure 6.2: Arc root contact time on fixed contact

6.2 Software computer programmes

A personal computer (PC) is used to control the sequence of the associated instruments. The computer (PC) is fitted with twin 16 channel input/output cards (I/O A and I/O B) with an onboard timer. The first I/O card is applied to control the AIS. The second I/O card is utilised to control the capacitor discharge bank, Flexible Test Apparatus (FTA) and also to trigger the Digital Storage Oscilloscope (DSO). There are three computer programs, all have been written in Qbasic computer programme. These programmes are to control the operation of the experimental equipment, analysing the experimental results and make a series of the arc images movies to observe the movement of the arc root from contact region into the arc chamber. Description of the programmes as follows:

6.2.1 Flexgav4.bas

This program is installed to run the test sequences operation. This program is used to control the sequences of the Flexible Test Apparatus (FTA), Arc Imaging System (AIS), the capacitor discharge system (CDS) and the digital storage oscilloscope (DSO). The software computer program Flexgav4.bas starts to run, the delay contact time is entered as required, the capacitor bank is then charged to the desired voltage, the apparatus operate sequentially.

The solenoid is fired and the test data is recorded in the RAM of the Arc Imaging System (AIS) as previously described [9,10,11,12,66]. Simultaneously the short circuit current and the short circuit voltage are displayed on the digital storage oscilloscope (DSO).

The experimental result is transferred from the memory of the digital storage oscilloscope (DSO) and from the Arc Imaging System (AIS) to the hard disk in the personal computer (PC). When this program is initiated, the Digital Storage Oscilloscope (DSO) will reset its system. A symbol READY status for recording a new test data acquisition is shown. A delay contact time (t_{cod}) is entered into the PC. After that the apparatus will start to charge the capacitor bank until the required voltage is reached.

6.2.2 Arcimage.bas

This program is able to present a series of images of the arc motion from the contact region until the arc reaches at the arc stack as a movie. It also replays the arc image contour movies.

6.2.3 Rootplot.bas

This program is used to analyse the experimental results. It shows cathode root contact time and anode root contact time separately with a facility to plot the arc root displacement. The arc power, arc voltage, arc current, etc also plot together with the arc root displacement.

6.3 Variable factors

To investigate the influence of Ag/C contact materials on the arc root motion from the contact region, a piece of Ag/C is welded on the fixed and the moving contact when the anode and cathode power supply are set in the fixed and moving contacts.

To inspect the influence of the short circuit current level on the arc root commutation from the contact region, three levels of the short circuit current are expected: 500 A, 1400 A and 2000 A. The Capacitor Discharge System (CDS) is charged up with capacitor 47.4 mF, inductor 224 μH and resistance 30 m² [2, 10-23].

To investigation the influence of the arc chamber materials on the arc root moving off from the contact region, a variety of arc chambers are created using interchangeable components. The top and bottom plates are plain flat rectangles of material. The plates are squeezed firmly into the assembly. The top pate is transparent to allow the high Speed Arc Imaging System (AIS) to record the arc motion in the arc chamber. The top plate is used the quartz glass and the bottom plate is used the polycarbonate or ceramic.

There are three vent plate configurations used in these experiments as shown in Figure 3.19 to investigate the influence of the arc chamber venting on the motion of the arc root commutes from the contact region. They are opened plate which has a vent area 40 mm², choked plate (vent area 15%) and closed plate which there is no vent area. The venting of the arc chamber is regulated by using a rectangular aperture with slot to hold a vent plate.

In all cases the methodology used is well established and involves using new materials in the test after 10 consecutive short circuit tests. There are 10-30 tests per condition repeated. The experimental results are presented with the error band of ±1 standard deviation.

6.4 Associated equipment operation

After the SPACEBAR is pressed the control computer energises the solenoid. The hammer then moves passes the optical sensor. This trigger sends a signal to the counter in the personal computer (PC) to commence a count down. The period from the hammer triggers the sensor and impacting the contact mechanism to execute the contact opening delay (t_{cod}) as required. The computer (PC) calculates the time period (t) from that period minus the contact opening delay time t_{cod}. This period (t) is loaded into I/O card to count down the time.

When the counter reaches zero, the trigger starts to produce the fault current from capacitor discharge bank (CDB). Then the hammer impacts the contact mechanism. The contact is opened and the arc is drawn. The data of the light intensity acquisition is recorded in the Arc Imaging System (AIS).

Then, the digital storage oscilloscope (DSO) records the fault current and the fault voltage profiles. The computer (PC) turns off the high power thyristor in the capacitor discharge bank. The sequentially Flexible Test Apparatus (FTA) is isolated from the capacitor discharge bank (CDB). Then the data stored on the memory of the Arc imaging System (AIS) and the Digital Storage Oscilloscope (DSO) is downloaded into the hard disk of computer (PC) via interface bus for further storage and analysis.

CHAPTER 7

EXPERIMENTAL RESULTS

7.1 Arc Current

This session show the experimental results of the arc current with short circuit current level 2000 A, cathode on the fixed contact and anode on the moving contact, ceramic arc chamber, choked arc chamber venting, contact opening velocity 1 m/s, 4 m/s, 5.5 m/s and 10 m/s.

The waveforms of the arc current are shown in Figure 7.1 to 7.4.

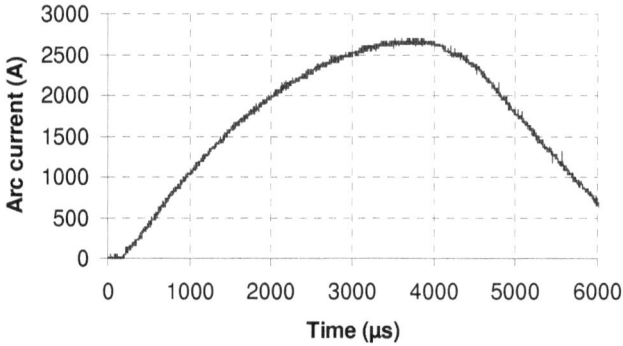

Figure 7.1: Arc current at contact opening velocity: 1 m/s.

The total period of the arc for the contact opening velocity at 1 m/s is longer than 6000 µs. This is the longest short circuit arc. The peak arc current level is about 2700 A. which is the highest compare to the arc current of contact opening velocity of 4 m/s, 5.5 m/s and 10 m/s

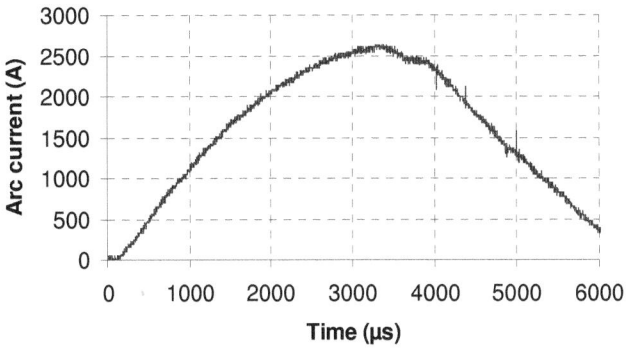

Figure 7.2: Arc current at contact opening velocity: 4 m/s

The total period of the arc for the contact opening velocity at 4 m/s is longer than 6000 μs but is shorter than the contact opening 1 m/s. The peak of arc current is about 2600 A which is higher than contact opening velocity 5.5 m/s and 10 m/s.

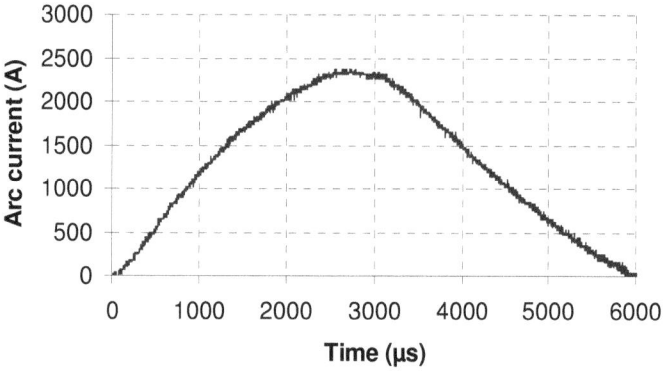

Figure 7.3: Arc current at contact opening velocity:
5.5 m/s

The total period of the arc for the contact opening velocity at 5.5 m/s is about 6000 µs. The peak of arc current is about 2400 A. This is higher than the arc current at contact opening velocity of 10 m/s.

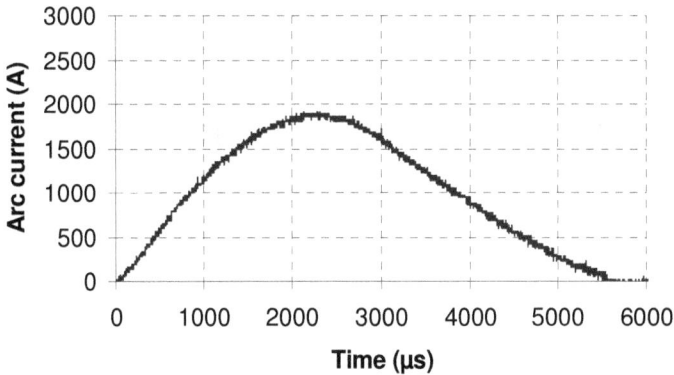

Figure 7.4: Arc current at contact opening velocity: 10 m/s

The total period of the arc for the contact opening velocity at 10 m/s is about 5500 µs. This is the shortest short circuit arc between 1 m/s, 4 m/s and 5.5 m/s. The peak of arc current is about 1900 A.

7.2 Arc Voltage

This session show the experimental results of the arc voltage with short circuit current level 2000 A, cathode on the fixed contact and anode on the moving contact, ceramic arc chamber, choked arc chamber venting, contact opening velocity 1 m/s, 4 m/s, 5.5 m/s and 10 m/s.

The waveforms of the arc voltage are shown in Figure 7.5 to 7.8.

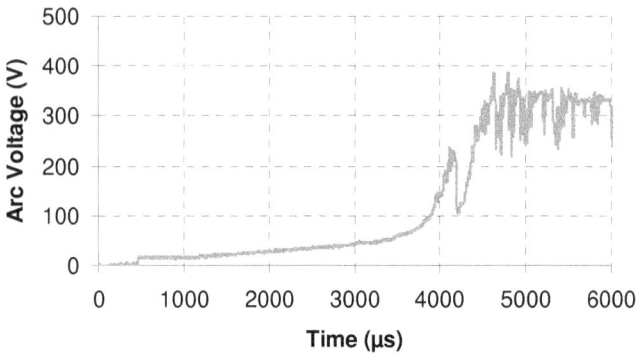

Figure 7.5: Arc voltage at contact opening velocity: 1 m/s

The arc root moves from the contact region into the arc stack at 4500 μs for contact opening 1 m/s. The period of the arc root moves from contact region into the arc stack for contact opening velocity of 1 m/s is longer than 4 m/s, 5.5 m/s and 10 m/s.

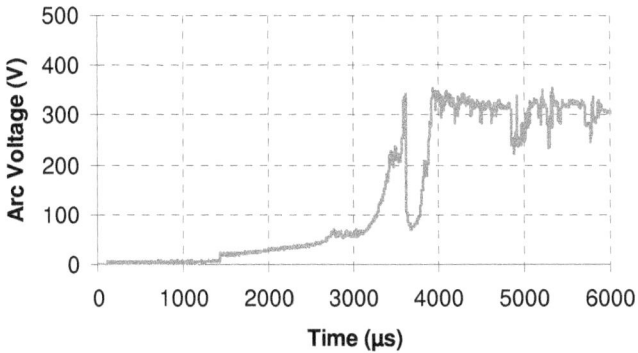

Figure 7.6: Arc voltage at contact opening velocity: 4 m/s

The arc root moves from the contact region into the arc stack at 4000 μs for contact opening 4 m/s. The profile of the arc voltage at the 3500 μs shows the arc reach the arc stack but return back to the arc chamber before moves into again at 4000 μs.

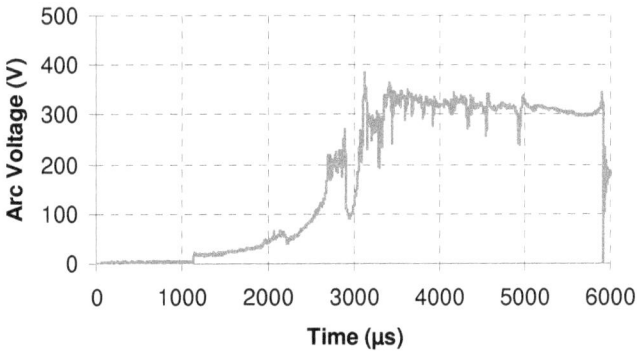

Figure 7.7: Arc voltage at contact opening velocity: 5.5 m/s

The arc root moves from the contact region into the arc stack at 3200 µs for contact opening 5.5 m/s. The waveform of the arc voltage of contact opening velocity 5.5 m/s shows the arc reaches the arc stack faster than the contact opening velocity of 1 m/s and 4 m/s but slower than the contact opening velocity of 10 m/s. The arc voltage comes down to normal level after get rid of the short circuit at 6000 µs.

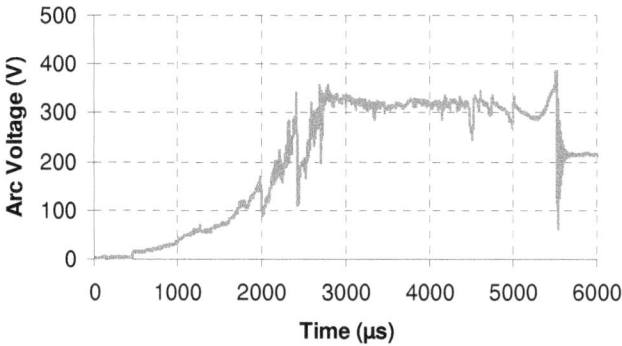

Figure 7.8: Arc voltage at contact opening velocity: 10 m/s

The arc root moves from the contact region into the arc stack at 2800 µs for contact opening 10 m/s. The waveform of the arc voltage of contact opening velocity 10 m/s shows the arc reaches the arc stack faster than contact opening velocity of 1 m/s, 4 m/s and 5.5 m/s. The arc voltage comes down to normal level after get rid of the short circuit at 5500 µs.

7.3 Arc Root Displacement

This session show the experimental results of the arc roots displacement (anode root and cathode root) with short circuit current level 2000 A, cathode on the fixed contact and anode on the moving contact, ceramic arc chamber, choked arc chamber venting, contact opening velocity 1 m/s, 4 m/s, 5.5 m/s and 10 m/s.

The waveforms of the, arc voltage, arc root: anode root and cathode root are shown in Figure 7.9 to 7.12.

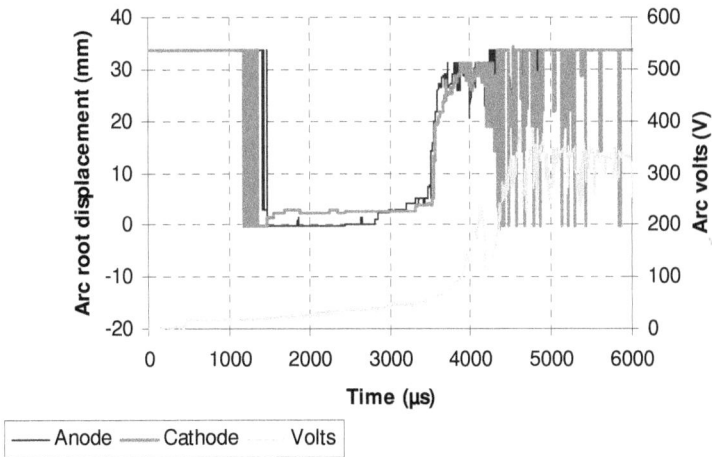

Figure 7.9: Arc roots displacement at contact
opening velocity: 1 m/s

The arc root delays to commute off from the
contact longer at contact opening velocity 1
m/s than that of 4 m/s, 5.5 m/s and 10 m/s.
The waveform of the arc root displacement
of contact opening velocity of 1 m/s shows
the arc root starts to leave the contact region
at 3500 μs.

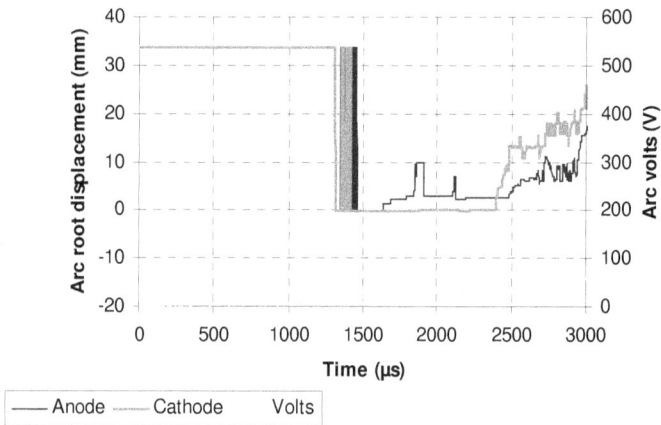

Figure 7.10: Arc roots displacement at contact opening velocity: 4 m/s

The waveform of the arc root displacement of contact opening velocity of 4 m/s shows the arc root starts to leave the contact region at 2400 μs. The arc root delays to commute off from the contact longer shorter than the arc root of contact opening velocity 1 m/s.

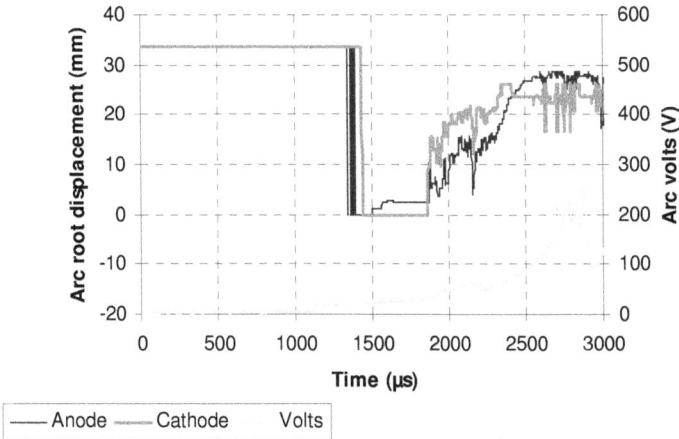

Figure 7.11: Arc roots displacement at contact
opening velocity: 5.5 m/s

The waveform of the arc root displacement
of contact opening velocity of 5.5 m/s shows
the arc root starts to leave the contact region
at 1800 µs. The arc root commute from the
contact faster than the arc root of contact
opening velocity 1 m/s and 4 m/s.

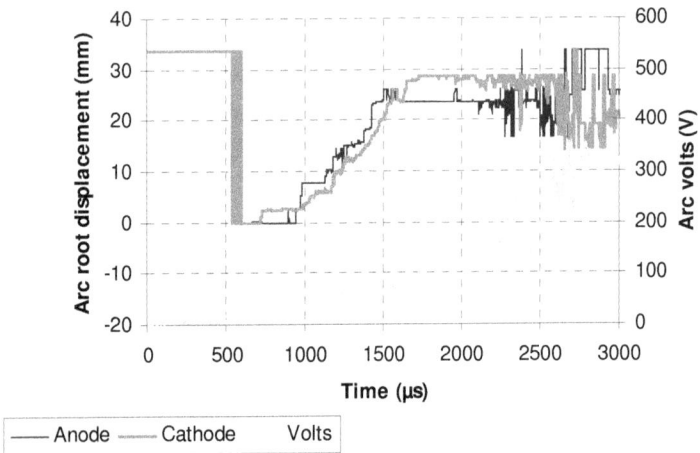

Figure 7.12: Arc displacement at contact opening
velocity: 10 m/s

The waveform of the arc root displacement
of contact opening velocity of 10 m/s shows
the arc root starts to leave the contact region
at 1000 μs. This arc root of the contact
opening velocity of 10 m/s is the fastest
commute from the contact compare to the
arc root of contact opening velocity 1 m/s, 4
m/s and 5 m/s.

7.4 Arc Pressure

The arc pressure profiles at contact opening velocity of 1 m/s were obtained from the pressure measurement are shown in figure 7.13 and 7.14.

The relationship between the arc root motion and differential pressure in the arc chamber is presented in Figure 7.13. The cathode and anode root move off from contact regions at 1940 μs (point "C") and 2670 μs (point "A"), respectively. The pressure measured here is referenced to ambient. The pressure in front and behind of the cathode arc root (at point "C") are 0.046 bar and 0.06 bar, respectively. At the point at which the anode root commutes from the moving contact (point "A"), the pressure in front the arc is 0.045 bar and the pressure behind the arc is 0.150 bar.

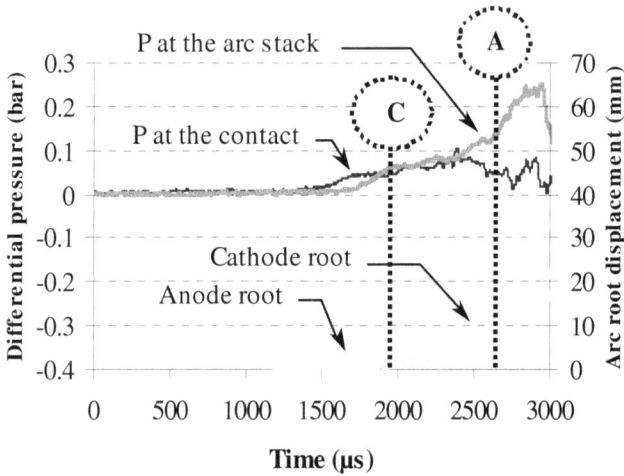

Figure 7.13:   Arc root displacement and differential pressure in the arc chamber, at contact velocity 1 m/s, polycarbonate arc chamber, Ag/C flat, Point "A" is the point at which the anode root moves from moving contact and point "C" is the point at which the cathode root moves from fixed contact [10]

The arc velocity increased as the ratio of pressure in the arc chamber increased as shown in figure 7.14. The arc velocity at the point at which the cathode moves from the fixed contact region is lower than the point at which the anode commutes from the moving contact region. The velocity of the arc increases rapidly when arc root commutes from the moving contact.

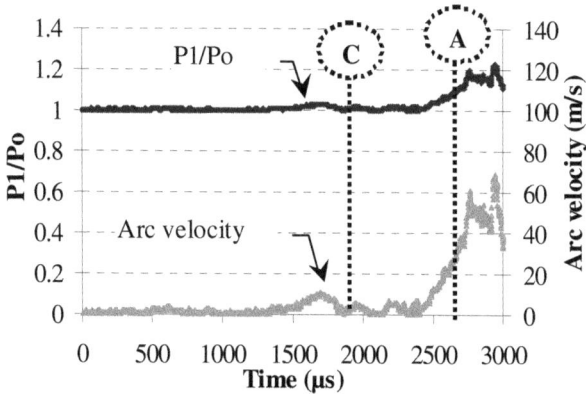

Figure 7.14: Arc velocity and pressure across the arc, Point "A" is the point at which the anode root moves from the moving contact and point "C" is the point at which the cathode root moves from fixed contact [10]

7.5 Piezoelectric Results

In Figure 7.15, the arc power does not vary strongly with contact opening velocity for the conditions shown. This suggests that the movement of the arc from the contact region is not a simple function of contact gap, but that the arc power is an important parameter in this process. From the results [3], the arc current increases as the contact opening velocity is decreased but the arc voltage decreases.

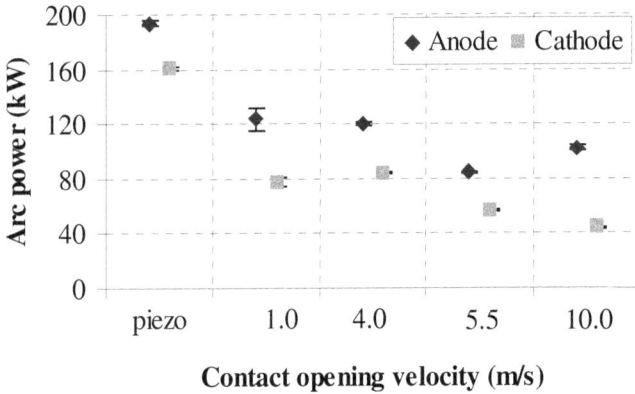

Figure 7.15: Arc power at the point that the arc root moves from the contact region and contact opening velocities [11, 18]

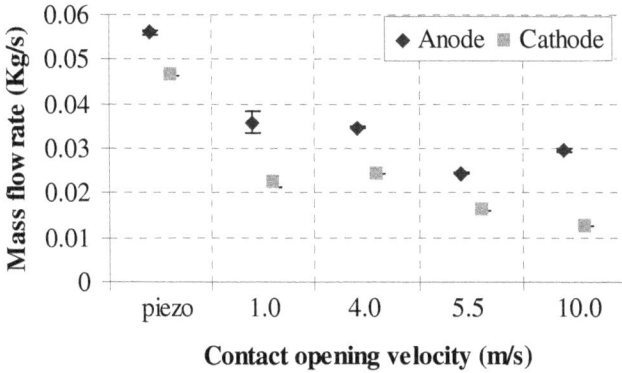

Figure 7.16: Mass flow rate at the point that the arc root moves from the contact region and contact opening velocities [11, 18]

The mass flow rate as shown in Figure 7.16, in the arc chamber can be estimated from the relationship between the arc power and the enthalpy from [3]. The arc root contact time with contact opening velocity is obtained from 1 m/s to 10 m/s [3]. Thus, the mass flow rate and the total mass flow rate at the point at which the arc root moves from the contact region as shown in Fig.7.16 and 7.17.

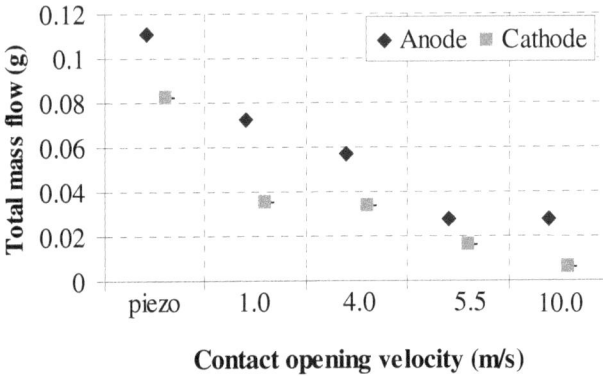

Figure 7.17: Total mass flow at the point that the arc root moves from the contact region and contact opening velocities [11, 18]

7.6 Arc Power and Displacement

The results of the arc root motion are shown in Figure 7.18, which can be viewed with the arc power and arc voltage data in Figure 7.19.

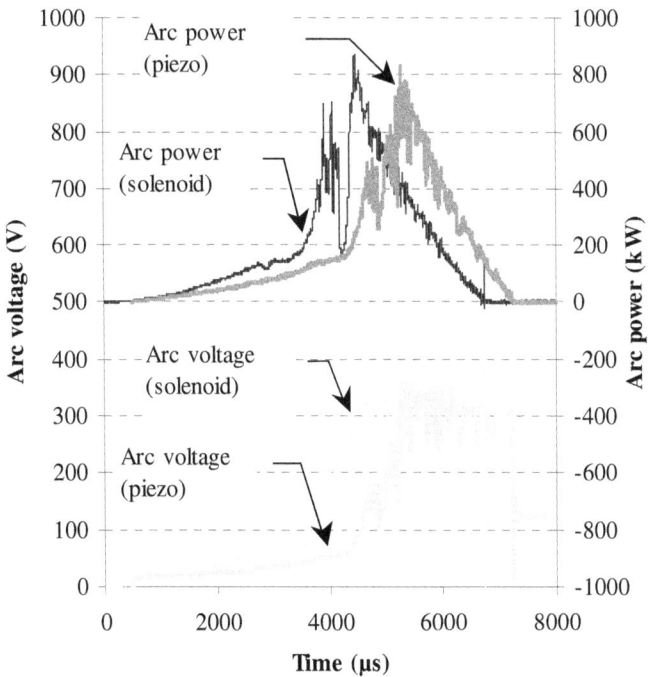

Figure 7.18: Arc voltage between piezo actuator and
1 m/s solenoid [18]

In Figure 7.18 the arc power and voltage
produced in the Flexible Test Apparatus is
compared with a result from [4] of the case
with a solenoid operated contact at 1 m/sec.
This shows that the arc voltage rise during
the short circuit event is similar indicating a
slightly lower contact velocity.

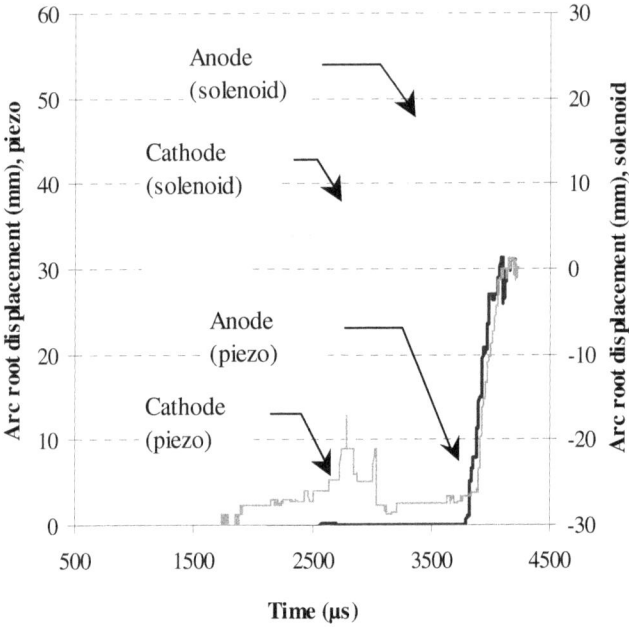

Figure 7.19: Arc root displacement with the PZT and (1 m/s) solenoid actuator [18]

The results of the arc root motion shown in Figure 7.19 show that the anode and cathode roots start to move at the same time. The cathode starts to move at 3.802 ms, and the anode (after 10mm displacement at 3.918 ms. while the results from the cathode starts to move at 2 ms anode at 3.24 ms.

7.7 Characterization of the PZT actuator

The PZT actuator is mounted such that top of the U-shape is fixed, the leg to which the dc supply is connected does not move, and the actuator movement can be measured at the tip of the moving leg. These results are shown below for a static demonstration of the device actuation.

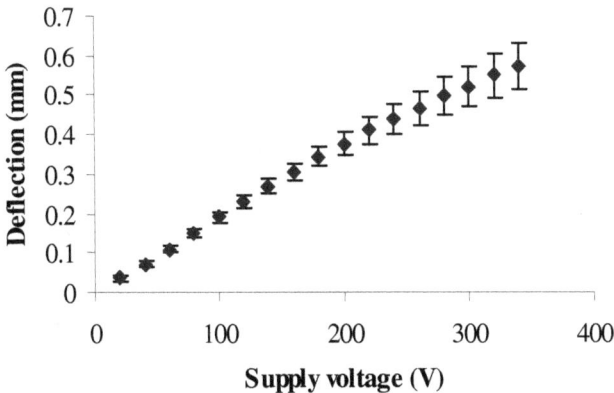

Figure 7.20: The relationship between piezoelectric supply voltage and deflection

The experimental results of the relationship between the supply voltage and the deflection of the piezoelectric are shown in Figure 7.20. The deflection of the piezoelectric increases linearity as the supply voltage is increased. At the maximum forward voltage at about 500 Vdc, the deflection of the piezoelectric shows less than 1 mm. From experimental results in [ref], the minimum contact gap that the arc root moves from the contact region is bigger than the maximum deflection of the piezoelectric.

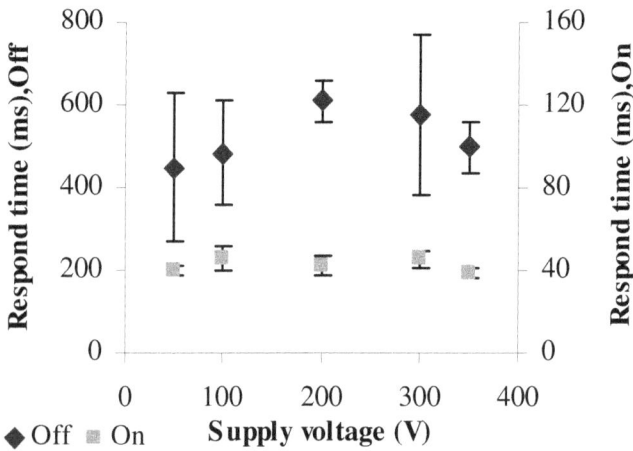

Figure 7.21: The respond time and supply voltage of piezoelectric at On and Off state

The respond time of the piezoelectric at the On state is about ten times faster than the off state, see Figure 7.21. The average of the respond time of the piezoelectric for the on state is stable as the supply voltage is increased. While in the off condition, the respond time has no significant on the supply voltage.

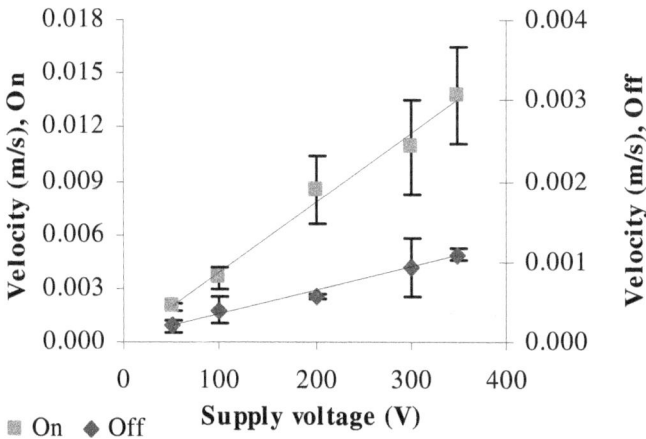

Figure 7.22: Deflection velocity and supply voltage of piezoelectric at On and Off state

The deflection velocity shows increasing as a linear as supply voltage of the piezoelectric is increased as shown in Figure 7.22. For the

"ON" condition the deflection shows higher velocity than the "OFF" condition at higher supply voltage

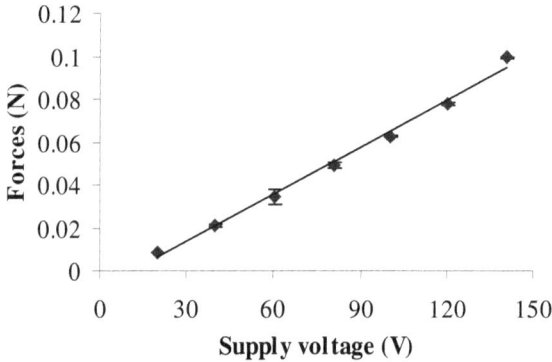

Figure 7.23: Forces on the piezoelectric and supply voltage

Figure 7.23 show the relationship between the forces on the piezoelectric and supply voltage. The results show that the forces on the piezoelectric increase linearity as the supply voltage is increased.

CHAPTER 8

DISCUSSION

Recent investigations [10-23] have shown that with an optimisation of the arc chamber, resulting from a detailed study of the arc-root phenomena, that the contact velocity can be reduced below the 6 m/sec requirement.

This now opens the possibility of refinement of the opening mechanism using smart materials actuation. The closer the match between actuator performance and the demands of the contact system, the simpler the operating mechanism becomes. The number of mechanical parts can be minimised, and the demands on their performance reduced. This will improve reliability, reduce size and power consumption and lower manufacturing costs. Piezo actuators are used in as the trip element in commercial circuit breakers [24].

Ultimately it may be possible to design a completely solid state contact system.

In this book the concept of peizo-ceramic actuation in current limiting circuit breakers is presented and discussed. A contact actuation system has been developed and is presented here. The performance of the contact system under short circuit conditions is analysed using the flexible test system and arc imaging system described previously [2, 10-23].

8.1 Piezoelectric Circuit Breakers

The piezo actuator can be used directly to open a set of contacts. With this particular actuator the peak velocity is below that normally required for effective current limitation.

However, recent research [10-23] has shown that it may be possible to improve arc control to an extent that contact velocities of the order of 1 m/s may be feasible thus permitting direct actuation. The deflection of the actuator is is in the right order of magnitude for a contact system, but marginal for a short circuit device. The force from the actuator would also be insufficient for direct contact action. For these reasons, the actuator needs to be used as a release for a higher force and displacement system. The discrepancy between the actuator performance and the requirements of the contact system needs to be minimised to provide the simplest mechanical system.

The ultimate goal of the research is to eliminate intermediate stages thus permitting direct contact actuation and the design of a solid state circuit breaker. It should be noted that a wide range of actuators with various performances are available and further optimisation of the actuator to the contact mechanism will be required.

8.2 Spring System

When applied in a circuit breaker mechanism the actuator behaviour can be approximated by a simple mass – spring system. The spring rate is independent of the applied voltage, but the unloaded position varies with voltage as shown in Fig.8.1.

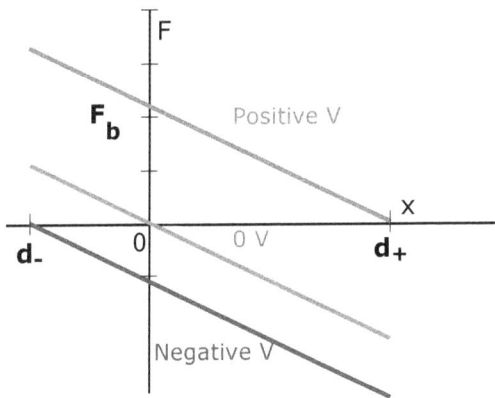

Figure 8.1: Spring model of piezo actuator.
d_+ is the free positive position, d_- the free negative position and F_b the blocked force

A negative voltage can be applied to consierably extend the operating range of the actuator up to a limit determined by the coercive field of the ceramic [34]. Only positive voltages are considered in this book.

The static spring characteristic of the actuator is described by the following equation relating the force, F, and deflection x:

$$F = k \left[d_0(V) - x \right]$$

where d_0 is the voltage dependent free displacement of the actuator and k is the spring constant. The blocked force, F_b is given by:

$$F_b = k \, d_0$$

The variation of free displacement with input voltage is shown in Figure 8.2. This particular actuator had a stroke of 1.2mm for the 400V 0.5Hz triangle wave used. The actuator stroke can be extended by electronic means as described above or by mechanical extension. The dynamic stroke will be considerably larger than this particularly at resonance.

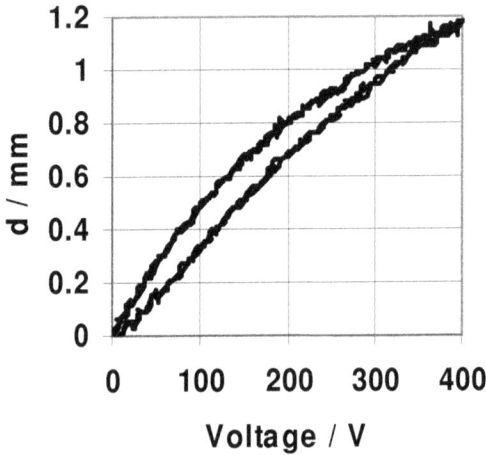

Figure 8.2: Voltage response of the planar bimorph actuator

In common with all piezo actuators, hysteresis and non-linearity are observed in the voltage response of the actuator [35] and this must be accounted for in the design.

The spring-like characteristic of the piezo actuator is different to that of the solenoid or electromagnet conventionally used in circuit breaker designs as described schematically in Figure 8.3.

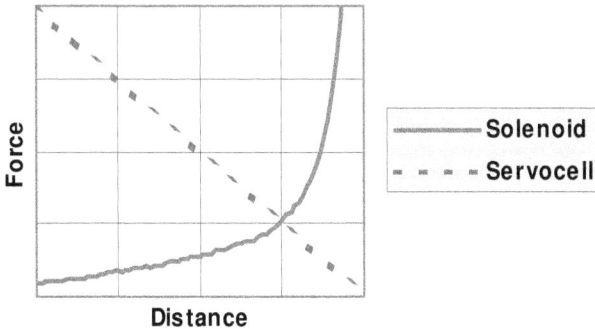

Figure 8.3: Force displacement profiles of piezo and solenoid actuators

In a circuit breaker the actuator will of course be coupled to a mechanism and this will play a large part in the dynamics of the system. To design the mechanism and predict its performance, the piezo can be built into a complete system model as a mass spring system. The spring rate and free deflection of the actuator are described above. The effective mass of the actuator can be obtained from the resonant frequency by assuming a simple harmonic oscillator:

$$\tau = 2\pi \sqrt{\frac{m}{k}}$$

$$m = \left(\frac{\tau}{2\pi}\right)^2 k$$

8.3 Arc Movement

The observation that the arc movement occurs at the same contact gap for both contact systems is in accord with previous studies [36] where it was shown that a minimum contact gap is requirement for arc motion. However, in this study the most significant delay in moving away from the contact region occurs after this initial arc motion, and in these tests the onset of rapid arc motion bore no relationship to the contact separation.

The waveform of arc voltage and contact position when the contact opening velocity of 1 m/s, using solenoid and piezoelectric as mechanism to open the contacts, as shown in Figure 8.4.

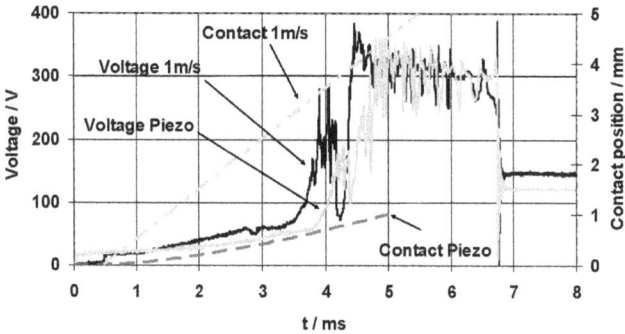

Figure 8.4: Arc voltages for piezoelectric and constant velocity 1 m/s contact systems. Contact trajectories for the two systems

8.3.1 Anode Root Movement

From figure 8.5, on the 1 m/s moving contact, the anode moves rapidly 2.5 mm after 1 m/s at a contact gap of 0.6 mm, after this movement the anode root remains static for 2ms until the arc transfers rapidly away from the contact region, at a contact gap of 2.5 mm.

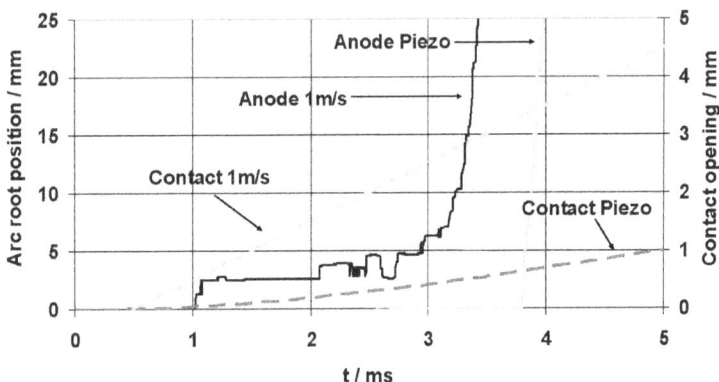

Figure 8.5: Anode root displacement compared between piezoelectric and constant velocity (1 m/s) contact systems, contact trajectories for the two systems

From figure 8.5, on the 1 m/s moving contact, the anode moves rapidly 2.5 mm after 1 m/s at a contact gap of 0.6 mm, after this movement the anode root remains static for 2ms until the arc transfers rapidly away from the contact region, at a contact gap of 2.5 mm.

On the significantly slower piezoelectrically operated contact the anode remains static for 3.8 ms until the contact gap reaches approximately the same value as for the initial movement in the constant velocity opening system (0.6 mm). The anode root is then able to transfer rapidly away from the contact system and into the arc stack.

8.3.2 Cathode Root Movement

In Figure 8.6, there is not a very large difference in the time of rapid arc motion despite significantly different contact velocities. In this case it is possible that the delay is caused by the transfer of the arc from the moving contact to the arc runner, where factors such as gas flow have a strong influence. The cathode root on the fixed contact shows a more gradual movement for both the piezoelectrically operated and constant velocity contact systems. The constant velocity system shows an initial slowarc rootmovementof approximately 5 ms followed by rapid arc motion at the same time as the anode.

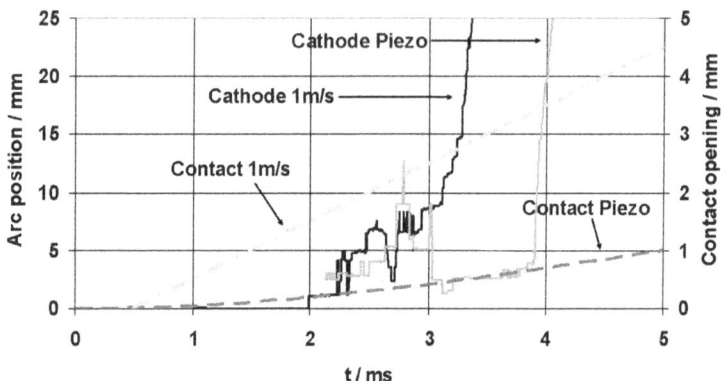

Figure 8.6: Cathode root displacement compared between piezoelectric and constant velocity 1 m/s contact systems. Contact trajectories for the two systems are shown.

The piezoelectric system shows some erratic movement between 2–3 ms. the onset of rapid arc motion occurs at the same contact gap as the start of the cathode movement in the constant velocity system. The effect of the contact opening velocity on arc erosion has not been investigated in this study, but the longer dwell time in the region of arc formation is likely to increase contact erosion.

8.4 Opportunity

The direct electromechanical coupling of a piezoelectric actuator means that it would be possible to tailor the contact opening trajectory simply by controlling the charge on the actuator. It is known that the opening velocity profile can have a significant effect on the contact erosion [36, 37], and it may therefore be possible to optimize the contact opening trajectory for minimum contact erosion. This is an important area for further study.

These initial results show that despite a low contact opening velocity on the piezoelectrically operated contacts (due to the large mass of the moving contact) the transition to rapid arc motion away from the contact region occurs only slightly later than the constant velocity system. Clearly the contact mass is a significant factor in the speed of operation of the contact system.

However, the overall effect on the arc control is not as significant as might be expected on the basis of a minimum contact gap requirement. Other factors such as optimization of the gas flow to permit rapid transfer off the moving contact are significant. Successful current limitation was therefore possible using a piezoelectric actuator. Even so there is considerable scope for improving the effectiveness of the piezoelectrically operated system not only by increasing the contact velocity but also by further development of the arc chamber to provide enhanced performance at low contact opening velocity.

CHAPTER 9

CONCLUSION

This book demonstrates how a piezoelectric actuator can be used as part of the actuation system in a circuit breaker mechanism. A methodology is presented for the analysis and design optimization of piezoelectric actuator based contact system. A piezoelectrically operated system was implemented in a test chamber and used to demonstrate successful current limiting performance [10-23].

- The performance of a commercially available piezoelectric ceramic actuator is assessed for suitability for use in a circuit breaker mechanism.

- The piezoelectric actuator is able to achieve contact velocities in the range required for current limitation for an arc chamber optimized for low contact opening velocity.

- The moving contact mass can significantly reduce the contact opening velocity and is therefore a significant design parameter.

- Short circuit tests show effective current limiting performance comparable to a constant velocity (1 m/s) contact system.

The relationship between the arc mobility and the contact opening velocity is complex and depends on the magnetic environment in the arc chamber and the flow of the arc chamber gases. The adverse gas flow effects can impede the arc root motion, and that optimization of the arc chamber design could permit the use of lower contact opening velocities.

The arc root phenomena can yield improvements in arc chamber performance thus permitting operation at reduced contact opening velocity. This now opens the possibility of refinement of the mechanism using smart materials actuation.

The design of circuit breaker systems could be simplified by the use of smart materials technology to provide the contact actuation.

The closer the match between actuator performance and the demands of the contact system, the simpler the operating mechanism becomes. The number of mechanical parts can be minimized, and the demands on their performance reduced.

This will improve reliability, reduce size, power consumption and manufacturing costs. As well as benefits in the design of conventional circuit breakers, the ability to make solid state contact opening mechanisms would allow substantial miniaturization of these devices resulting in a completely new type of device.

References

1. R.T. Lythall, "The switchgear book", Butterworth & Co., Ltd, London, 1972.

2. P.A. Jeffery," The motion of short circuit arcs in low current limiting miniature circuit breakers", Thesis submitted for PhD, University of Southampton, Jan. 1999.

3. Warren B Boast, "Vector fields", Harper & row, London, UK, 1964.

4. J.M.Somerville, "The Electric Arc", Methuen & Co., LTD, London, 1959.

5. Cassie A.M., "Power System Switchgear-The Physical Nature of Arcs", The electrical Journal 3, April 1959.

6. Paul G. Slade, "Electrical contacts Principles and Applications", Marcel Dekker, inc., New York, 1999.

7. F. Edlmayr, K. Krause, S. Theodor, " Low voltage switchgear", Heyden & Sons Ltd., London, 1973.

8. J.S. Morton, Air-breaker circuit breaker, Power Circuit breaker Theory and design, Institution of Electrical Engineering, England, 1975, pp.185-187.

9. P. M. Weaver, J. W. McBride, "Arc motion in current limiting circuit breakers", 16 International conference on Electrical Contacts, Sep 1992, UK, pp.285-288

10. Kesorn Pechrach, Textbook "Arc Control in Circuit Breakers", ISBN: 978-3-639-22101-5, 316 pages, 21 December 2009, VDM Publishing house Ltd, Germany.

11. P.M.Weaver, K.Pechrach, J.W.McBride, "Arc Root Mobility on Piezoelectrically Actuated Contacts in Miniature Circuit Breakers", IEEE Transactions on Components, Packaging, and Manufacturing

Technologies, Vol. 28, No.4, December 2005, pp. 734-740.

12. P.M.Weaver, K.Pechrach, J.W.McBride, "The Energetics of Gas Flow and Contact Erosion During Short Circuit Arcing", IEEE Transactions on Components and Packaging Technologies, Vol. 27, No.1, March 2004, pp. 51-56.

13. J.W.McBride, K.Pechrach, P.M.Weaver, "Arc Motion Gas Flow in Current Limiting Circuit Breakers Operating with a Low Contact Switching Velocity", IEEE Transactions on Components, Packaging, and Manufacturing Technologies, Vol. 25, No.3, September 2002, pp. 427-433.

14. J.W.McBride, K.Pechrach, P.M.Weaver, "Arc Root Commutation from Moving Contacts in Low Voltage Devices", IEEE Transactions on Components and Packaging Technologies, Vol. 24, No.3, September 2001, pp. 331-336.

15. K. Pechrach, "Piezoelectric Force Microscopy (PFM) Materials Characteristics for Medical Application", ATPER 2011, Copenhagen, Denmark, May 2011.

16. K. Pechrach, "Arc Control in Circuit Breakers with low Contact Velocity", TCP2010 Green Thailand, Bangkok, Thailand, July 2010.

17. K. Pechrach, "Electroceramics Green Energy Harvesting in Industrial Estates and Agriculture in Thailand", 12th ATPER 2009 - The Association of Thai Professionals in Europe, Paris, France, May 2009.

18. K. Pechrach, J.W. McBride, P.M. Weaver, Arc root mobility on piezo actuated contacts in miniature circuit breakers, 49th IEEE Holm Conference on Electrical Contacts, Washington DC, USA, September 2003.

19. K.Pechrach, J.W.McBride, P.M.Weaver, "The Correlation of Magnetic, Gas dynamic and Thermal Effects on Arc Mobility in Low

Contact Velocity Circuit Breakers", 48th IEEE Holm Conference on Electrical Contacts, Florida, USA, October 2002, pp.86-94.

20. K.Pechrach, J.W.McBride, P.M.Weaver, "Analysis of Arc root Mobility in Low Contact Opening Velocity Circuit Breakers", 21st ICEC/ITK 2002 International Conference on Electrical Contacts, Zurich, Switzerland, September 2002, pp.260-267.

21. K.Pechrach, J.W.McBride, P.M.Weaver, "Gas Flow and Composition Effects on Arc Motion in Current Limiting Circuit Breakers", 47th IEEE Holm Conference on Electrical Contacts, Montreal, Canada, September 2001, pp.12-17.

22. J.W.McBride, K.Pechrach, P.M.Weaver, "Arc Root Commutation from Moving Contacts in Low Voltage Devices", 46th IEEE Holm Conference on Electrical Contacts, Piscataway, NJ, USA, September 2000, pp.130-138.

23. J. W. McBride and P. M. Weaver, "Review of arcing phenomena in low voltage current limiting circuit breakers," Proc. Inst. Elect. Eng., vol. 148, no. 1, pp. 1–7, Jan. 2001.

24. S. C. Powell, "The development of a low cost, high performance piezo actuator for high volume manufacture," in Proc. 7th Int. Conf. New Actuators (Actuator'00), Bremen, Germany, Jun. 2000, pp. 45–48.

25. P. M. Weaver, S.C. Powell, "LOW COST PRECISION CONTROL OF PIEZO-CERAMIC ACTUATORS", Actuator 2002, Proc. 8th International Conference on New Actuators, Bremen, 10-12 June 2002,

26. Newcomb C.V., Flinn I. "Improving the linearity of piezoelectric actuators" Electronics Letters v.18 n.11 May 1982 pp.442-444

27. P.M Weaver, Y. Zheng, "THERMAL RESPONSE OF LARGE MOVEMENT LOW

POWER DC ACTUATORS FOR USE IN LOCK AND VALVE MECHANISMS", Actuator 2004, Proc. 9th International Conference on New Actuators, Bremen, 14-16 June 2004,

28. P.M. Weaver, A. Ashwell, Y. Zheng, S.C. Powell, "Extended temperature range piezo actuator system with very large movement" SPIE Smart Structure and Materials conference: Active Materials: Behaviour and Mechanics, San Diego, March 2003, pp.484-492.

29. M.W. Hooker, "Properties of PZT based piezolelectric ceramics between -150 and 250°C" NASA report: NASA/CR 1998-208708, 1998

30. M. Stewart, M. Cain "Piezoelectric Resonance" NPL GPG No.33 2001

31. P.M. Weaver, F. Graser, "A SENSORLESS DRIVE SYSTEM FOR CONTROLLING PIEZOELECTRIC

ACTUATOR HYSTERESIS", Actuator 2006, Proc. 10th International Conference on New Actuators, Bremen, 14-16 June 2006,

32. UK patent application No.0602255.

33. A. J. Moulson and J. M. Herbert, Electroceramics. London, U.K.: Chapman and Hall, 1990.

34. W. Rieder, "Interaction between magnet-blast arcs and contacts," in Proc. 28th Holm Conf. Electrical Contacts, 1982, pp. 3–10.

35. J. W. McBride and M. A. Sharkh, "The effect of contact opening velocity and the moment of contact opening on the AC erosion of Ag/CdO contacts," IEEE Trans. Comp., Packag., Manufact. Technol. A, vol. 17, no. 1, pp. 2–7, Mar. 1994.

36. E. M. Belbel and M. Lauraire, "Behavior of switching arc in low-voltage limiter circuit breaker," IEEE Trans. Comp., Hybrids, Manufact. Technol., vol. 8, no. 1, pp. 3–12, Mar. 1985.

37. B. Jaffe,W. R. Cook, and H. Jaffe, Piezoelectric Ceramics. New York: Academic, 1971.

38. Mark Stewart, Markys Cain and Paul Weaver, Use of Piezoceramics as DC Actuators in Harsh Environments, Actuator 2008, Proc. 11th International Conference on New Actuators, Bremen, 9-11 June 2008,pp. 67-70.

39. C. Liu, "Development of Surface Micromachined Magnetic Actuators usingElectroplated Permalloy," Journal ofMechatronics, pp.613-633, 1998.

40. B. Wagner, W. Benecke, G. Engelmann and J. Simon, "Micro actuators with moving magnets for linear, torsional or multi-axial motion," Sensors and Actuators, A(32), pp. 598-603, 1992.

41. C. Bolzmacher, K. Bauer1, R. Wagner, C. Schmutz, A. Gerber, E. Quandt, "Electromagnetic Microactuators Based on

Thin Solenoidal and Toroidal Coils", Actuator 2008, Proc. 11th International Conference on New Actuators, Bremen, 9-11 June 2008, pp. 329-332.

42. N. Kobayashi, A. Urakawa, T. Nanataki, "Low-Cost Piezoelectric Actuators with Honeycomb Structure", Actuator 2008, Proc. 11th International Conference on New Actuators, Bremen, 9-11 June 2008, pp. 531-533

43. Internet website http://www.ngk.co.jp/

44. K. Schmidt et. Al. Aktive Schwingungskompensation an einer PKW-Dachstruktur Adaptronic Congress, Wolfsburg (2003)

45. T. Pfeiffer, J. Nuffer, Reliability investigation of an automobil oilpan equipped with active noise reduction, DVM-Arbeitskreis Zuverlässigkeit mechatronischer und adaptronischer Systeme, Koblenz (2008)

46. E. H. Anderson, B. Houghton, Elite-3 Active Vibration Isolation Workstation, Proceedings of SPIE Volume 4332-23 Smart Structures and Materials, Newport Beach, CA, USA (2001)

47. M. Matthias, T. Melz, M. Thomaier, Entwicklung, Bau und Test eines multiaxialen, modularen Interfaces zur aktiven Schwingungsreduktion für Automotive Anwendungen" (in German), Adaptronic Congress Göttingen (2005)

48. M. Busse, F.-J. Wöstmann, T. Müller, T. Melz, P. Spies, Intelligent Cast Parts - Application of adaptronic Components with Cast Parts. Adaptronic Congress Göttingen (2006)

49. V. Bräutigam, P. Wierach, C. Körner, R. F. Singer, Integration of Piezoceramic Modules in Die Casting - A New Production Technology. Adaptronic Congress Göttingen (2006)

50. V. Bräutigam, C. Körner, R. F. Singer, M. Kaltenbacher, M. Meiler, R. Lerch: Smart structural components by integration of sensor/actuatormodules in die castings, Proceedings of SPIE Volume 6527 Int. Symposium of Smart Structures and Materials & Nondestructive Evaluation and Health Monitoring, 18 - 22 March 2007, San Diego, CA.

51. T. Pfeiffer, J. Nuffer, "Thermal Degradation of Piezoceramic Actuators due to Integration into Adaptronic Components by Die-Casting", Actuator 2008, Proc. 11th International Conference on New Actuators, Bremen, 9-11 June 2008, pp. 534-537.

52. A. Kappel, B. Gottlieb, C. Wallenhauer, Roland Zeichfüßl, "PAD - A Scalable Drive Technology", Actuator 2008, Proc. 11th International Conference on New Actuators, Bremen, 9-11 June 2008, pp. 558-561.

53. R.W. Jones, "Hysteresis Compensation in Smart Actuators: A Survey", Actuator 2008,

Proc. 11th International Conference on New Actuators, Bremen, 9-11 June 2008, pp. 935-940.

54. R. Smith, Smart Material Systems: Model Development. SIAM (2005)

55. I.D. Mayergoyz, Mathematical Models of Hysteresis and Their Applications. New York, Elsevier (2003)

56. M. Brokate, J. Sprekels, Hysteresis and Phase Transitions, New York, Springer-Verlag (1996).

57. S. Salapaka, A. Sebastian, J.P. Cleveland, M.V. Salapaka, Review of Scientific Instruments, 73(9), 3232-3241 (2002)

58. G. Schitter, P. Menold, H.F. Knapp, F. Allgower, A. Stemmer, Review of Scientific Instruments, 72(8), 3320-3327 (2001)

59. K. Kuhnen1), H. Janocha1), D. Thull2) and A. Kugi2), "A NEW DRIVE CONCEPT FOR HIGH-SPEED

POSITIONING OF PIEZOELECTRIC ACTUATORS ", Actuator 2006, Proc. 10th International Conference on New Actuators, Bremen, 14-16 June 2006, pp. 82-85.

60. A. Wolff*, C. Schuh*, K. Lubitz**, T. Steinkopff*, D. Jozinovic, "INTRODUCTION OF NOVEL SINGLE LAYER PIEZOELECTRIC MICRO ACTUATOR WITH GIANT DEFLECTION", Actuator 2006, Proc. 10th International Conference on New Actuators, Bremen, 14-16 June 2006, pp. 74-77.

61. Lauric GARBUIO, Jean-François ROUCHON, "PIEZOELECTRIC THRUST BEARING FOR SEVERE ENVIRONMENTS", Actuator 2006, Proc. 10th International Conference on New Actuators, Bremen, 14-16 June 2006, pp. 185-188.

62. H. H. Gatzen1, M. Hahn1, M. Bedenbecker1, B. Ponick2, R. Gehrking2, and S. Demmig2, "ADVANCES IN THE

DEVELOPMENT OF A LINEAR HYBRID MICRO ACTUATOR", Actuator 2006, Proc. 10th International Conference on New Actuators, Bremen, 14-16 June 2006, pp. 207-210.

63. Gatzen, H.H., Stoelting, H.D., Ponick, B.: 9th Int. Conf. On New Actuators, Proceedings Actuator 2004, Bremen, pp. 317-320.

64. Budde, T., Gatzen, H.H.: Proceedings, MMM, San Jose 2006.

Index